VICTORIAN SCIENCE AND ENGINEERING

PORTRAYED

In The Illustrated London News

The Illustrated London News *catered for a wide variety of interests. These included science, which it supported by means of a special column and by reporting and illustrating activities at the Royal Institution and Royal Polytechnic. This latter institution was renowned for spectacular demonstrations of popular science. Here its director, 'Professor' Pepper, is shown with the giant induction coil, capable of producing a spark, 'or rather a flash of lightning', 29 inches long.*

Victorian Science and Engineering Portrayed *In The Illustrated London News*

Kenneth Chew and Anthony Wilson

SCIENCE MUSEUM

ALAN SUTTON

First published in the United Kingdom in 1993 by
Alan Sutton Publishing Limited
Phoenix Mill · Far Thrupp · Stroud · Gloucestershire
in association with the Science Museum

First published in the United States of America in 1993 by
Alan Sutton Publishing Inc · 83 Washington Street · Dover · NH 03820

British Library Cataloguing in Publication Data

Chew, Kenneth
Victorian Science and Engineering
I. Title II. Wilson, Anthony
608.741

ISBN 0–7509–0326–0

Library of Congress Cataloging in Publication Data applied for

Typeset in 10/13 Sabon.
Typesetting and origination by
Alan Sutton Publishing Limited.
Printed in Great Britain by
Redwood Books, Trowbridge, Wiltshire

Contents

Acknowledgements *vi*

Foreword *vii*

Introduction *ix*

Food, Work and Health *1*

Communications and Transport *35*

Science: Created, Explained, Applied and Exhibited *91*

War *125*

Sources *142*

Bibliography *144*

Index *147*

Acknowledgements

We are glad to acknowledge the help we have received from our colleagues in the Science Museum Library, from Di Hunter, from Margaret Wilson, and from curators at the Science Museum and National Railway Museum who read the text and helped us to avoid a number of pitfalls; remaining imperfections are of course all our own work. Generous advice and encouragement were provided by the present Editor-in-Chief, *Illustrated London News Publications*, Mr James Bishop.

We are grateful to all these people, and especially to our Science Museum photographer, John Lepine, whose essential contribution to this book is manifest on almost every page. All the photographs are from volumes in the Science Museum Library.

We are grateful also to Thames and Hudson Ltd for permission to quote from *The History of Photography* by H. Gernsheim (© Helmut and Alison Gernsheim, 1969).

Kenneth Chew, Research Fellow,
Science Museum
Anthony Wilson, Publications Manager,
Science Museum
March 1993

Foreword

In the issue of *The Illustrated London News* published on 14 May 1892 to celebrate its fiftieth anniversary the then Art Editor, Mason Jackson, looked back on his long career in 'Pictorial Journalism': '. . . nothing attracts the multitude like *news*. Let the Queen or the Prince of Wales perform some great public function, which is promptly and fully illustrated in the Pictorial Press, and the printing machines cannot go fast enough to supply the demand. Let there be a storm, a disastrous fire, a fatal railway or mining accident, which excites the sympathy of the multitude, and the sale of the illustrated newspaper leaps up by thousands.' But it would be wrong to imagine that *The Illustrated London News* made its reputation on cheap sensationalism. On the contrary, what it aimed to provide, for a new, expanding and increasingly affluent middle-class market, was a comprehensive view of the world at large at a time when that world was changing as never before.

At the root of much of that change was the extraordinary progress being made in science, invention and engineering. Its effects were widely seen in Britain, not only because of the transformation that had taken place in British society and the national economy as a result of what was now being called the Industrial Revolution, but through the spread of that wealth of engineering expertise and of manufactured goods throughout the Empire and the newly industrializing world beyond. *The Illustrated London News* reflected these developments in its unique pictorial record of events and achievements. As Mason Jackson reflected, 'In looking down through the long vista, I see the triumphs of science in the girdle that has been put round the earth by the electric telegraph – in the telephone, that enables us to speak to our friends when miles away – in the wondrous invention by which we can even listen to the voice of the dead. I see the marvels of engineering skill in the mountains that have been tunnelled and in the rivers that have been spanned by stupendous bridges. I see the earth yielding up new and unsuspected treasures of gold and diamonds and fountains of oil for the use and enrichment of man.'

The Victorian age was one of supreme self confidence. Its heroes were the men of science and engineering whose achievements, almost without exception, were seen to have delivered benefits, immeasurable in their magnitude, to the nation and its people. They had created the wealth on which that self confidence was founded, and on which the ideals and values of enlightenment and democracy rested. It was expected and accepted that their great works should form a continuing thread through the pages of *The Illustrated London News*. It is fitting too that the Science Musum should co-publish a selection drawn from the Victorian years. For not only is the Museum itself a product of Victorian beliefs and values, but the core of its outstanding collections reflects that same faith in science and engineering. 'None can overestimate the teaching of the eye,' stated the Jubilee editorial; that sentiment applies equally to the collections of the Science Museum.

Neil Cossons
Director
National Museum of Science & Industry
March 1993

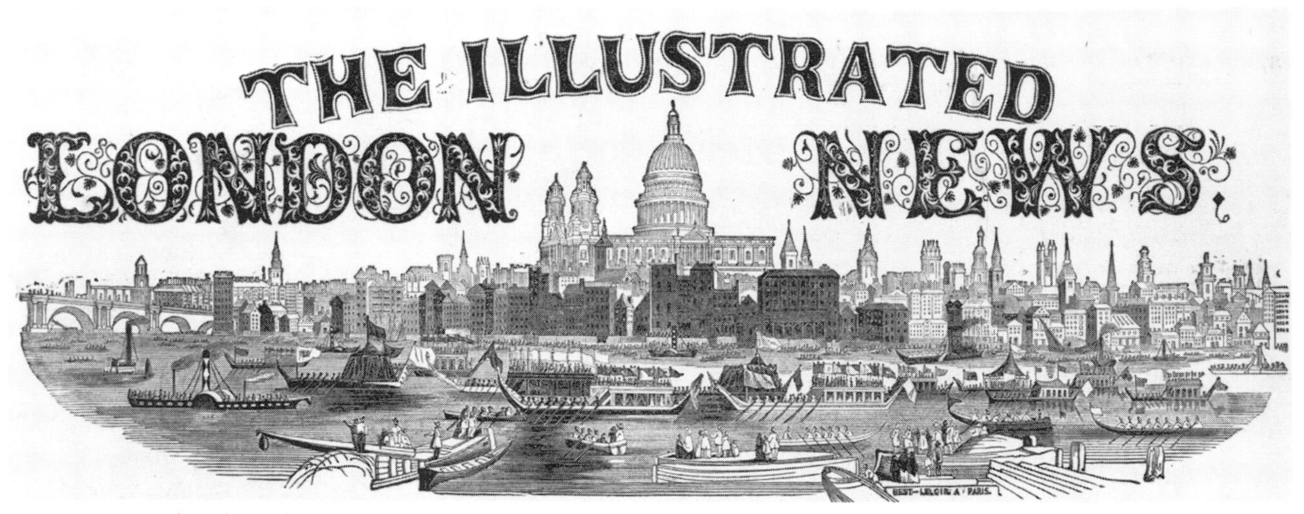

Mast-head used by The Illustrated London News *in 1850*

Introduction

In 1992 *The Illustrated London News* (*ILN*) celebrated 150 years of continuous publication. During that period literally thousands of its illustrations were related, directly or indirectly, to science and engineering. The present book contains reproductions of about one hundred of these engravings, selected from those which appeared between 1842 and 1901, i.e. within the Victorian era. Qualifications for the selection of a particular illustration were that it should be of a high artistic and technical quality, and that we should feel it to be, in one way or another, striking. The title of the selection, *Victorian Science and Engineering*, is a short title; a fuller one would have involved Science, Engineering, Technology, Industry, Agriculture and Medicine, these being the general fields covered in this book and in the Science Museum, London, with which both authors are associated. Within these fields the Museum has well over one hundred specific areas of interest, from Acoustics to X-rays, and it was decided that a striking illustration, from home or abroad, would be eligible if related to one or other of these areas.

In making the selection we followed no definite story line, nor did we attempt to make the text a condensed history of science and engineering, illustrated by *ILN* pictures. Nevertheless the selected pictures fell into groups within which they were related to one another, the short titles for these groups being *Food, Work and Health, Communications and Transport, Science: Created, Explained, Applied and Exhibited,* and *War*. The last of these may at first sight cause surprise, but preparations for war, and war itself, are notable accelerators of technical change, and they do produce some dramatic pictures.

What made a picture in these groups striking to us? Three of many possible examples will be considered. We were likely to find a picture striking if it represented the successful work of one of the great Victorian engineers, for not only was the subject likely to be spectacular, but also the artist was likely to have been inspired both by the subject itself and by the legitimate patriotic pride to which it gave rise. Such pictures, of Brunel's *Great Britain*, of Paxton's Crystal Palace, of Fowler and Baker's Forth Bridge, for example, are therefore strongly represented here. If allowed to do so they could have monopolized the book.

Secondly a picture could be striking even if it represented failure rather than success. A reader who looks up the Tay (railway) Bridge in the index would probably be seeking a picture, not of the present structure which has stood honourably since 1887, but of the earlier one by Bouch, part of which fell, carrying with it a train, on the night of 28 December 1879.

Herbert Ingram (1811–60), founder of The Illustrated London News

Thirdly we might find a picture striking if it had human as well as technical interest. Our best example of this is no. 87, in which an excellent portrait of a great American inventor is combined with an accurate delineation of his favourite invention. The impact of a picture of this kind might be further enhanced because we knew more about the subject and his destiny than did the artist. One such picture is no. 66 which shows two portly gentlemen watching two boys learning to ride the recently invented boneshaker in 1869. Interest thickened when we realized that one of the gentlemen was the French emperor and that one of the boys was his son, the Prince Imperial. But we knew what the artist could not know – that two years later the emperor would be in exile after defeat in war, and that ten years later the Prince Imperial would be killed by Zulu spears.

The text accompanying each picture is intended to enhance the interest created by it and to satisfy as far as possible any curiosity that it arouses. In some cases the text gives a commentary on the picture; in others it places the subject into its technological or social context; in others human interest is stressed. In many cases a reference to a book that can satisfy more fully the interest created can be found in the list of sources and bibliography on pages 142–6.

If this book has a hero it is Herbert Ingram, founder of the *ILN*. Born in 1811 in Boston, Lincolnshire, he completed his formal education in 1825, but apprenticeship to a printer provided for his lively mind a substitute for secondary and tertiary education. After a short period in London he established with his brother-in-law in 1833 a newsagency and printing business in Nottingham. Becoming aware that illustration, even on a small scale, would enhance the sale of periodicals, he moved back to London with the idea of a fully illustrated weekly newspaper in mind. All went well for him: he was well advised, he acquired by chance an able editor, and a high quality of illustration was established when eight of the twenty illustrations in the first number, of 18 May 1842, were supplied by a young man who eventually became Sir John Gilbert, RA. The *ILN* flourished and Ingram prospered, becoming MP for Boston in 1856. But in 1860, at the age of forty-nine, he was drowned with his son Herbert in an accident on Lake Michigan.

Most of our illustrations are boxwood engravings, made by a technique which was introduced by Thomas Bewick (1753–1828), and which dominated illustration until the late 1880s. The artist made the drawing

Printing The Illustrated London News: *Hoe's American Machine (left) and Ingram's Rotary Machine, 1879*

(possibly copied from a photograph) with pencil or pen and ink on the prepared end-grain surface of a small boxwood block 7/8 of an inch thick, this being the thickness of letter-press type. Alternatively an image might be directly photographed onto a suitably sensitized block. The block then passed to the engraver, who might be the artist himself, or a freelance, or a member of a studio of engravers. Large illustrations might need up to twenty blocks (possibly prepared by different engravers) clamped together. Sometimes the clamping was not entirely successful and the final print was criss-crossed by straight lines. Alternatives to the wood block began to appear in the 1880s (with a very pronounced drop in quality of the product), and by 1895 the *ILN* was using rotogravure, a process suitable for rotary presses, in which the plates were made by a combination of photography and etching. (For a fascinating account of the above processes see de Maré (1890).)

Here then is our gallery of illustrations. We hope readers will get as much pleasure from examining them as we did from collecting them.

Queen Victoria opens a new wing of the London Hospital, 1876

Food, Work and Health

Three opening pictures relating to food show respectively an early and unsuccessful attempt by technology to make the horse redundant in ploughing, a palace for the display of the results of cattle breeding, and a steamship in dock with its cargo of tea from China.

The human need for shelter is here symbolized by the city, the growth of which was such a conspicuous feature of the Victorian age. What better city could be chosen than the one described as 'the best designed Victorian town in England'?

One of the newer forms of lighting is inevitably illustrated by a gas holder, one of those aggressive structures that lowered the rents on the leeward sides of towns. Electric light is illustrated by rather a late example (1890), namely its introduction into the British Museum galleries.

Men and women at work are shown in a locomotive factory, in a steelworks, in a woollen factory and elsewhere. We show coals from Newcastle in London Docks, and one only of the tragically many pictures of the price paid in human lives for cheap coal. The monotony of much factory work is suggested by a picture of women slitting pen nibs in Birmingham.

Public enterprise in Victorian times is represented here by the building of London's Thames Embankment and the damming of the Vyrnwy valley to provide water for the city of Liverpool. Two aspects of Victorian medicine, namely the development of aseptic surgery, and the establishment of nursing as a profession, are represented by the appropriate portraits, and royal patronage of hospitals is pictured in the frontispiece. Two good portraits enable us to pay tribute to the work abroad of Robert Koch and Louis Pasteur.

1 James Boydell's steam horse, 1857

Such was the interest raised when James Boydell first exhibited his steam horse in a Lincolnshire field that spectators from as far afield as Cuba and Russia were among the crowd. Boydell was seeking the fame and fortune that would come to the first person to harness steam power successfully to cultivation. The steam engine had already begun to take over static tasks on the farm, such as threshing, but in the fields it was too cumbersome to be practical. Boydell's novel proposal was the 'endless rail', a series of hinged boards attached round the perimeter of each wheel; like a skier on soft snow, his steam horse would spread its weight and thus avoid becoming bogged down.

The *ILN* correspondent was impressed by the demonstration. With three double ploughs in tow, the engine could plough an acre in little over an hour. At the end of the day it clanked off up a steep incline pulling two heavy wagons; Boydell was rewarded with three hearty cheers, and his men shared £4 collected from bystanders.

The 'endless rail' enjoyed a modicum of sucess in the following years, but it was not the solution to the problem. The hinged boards were easily damaged and hard to maintain, and the engine still tended to compact the ground unacceptably. Mobile steam power did eventually offer some help with ploughing, but only when the engine itself remained on the sidelines, manoeuvring a plough across the field by a system of cables and pulleys.

2 The Smithfield Cattle Show, 1862

The prosperous yeomanry of Victorian Britain came in tens of thousands to the annual Smithfield Cattle Show in London. Cabbies waited at the main-line termini to pounce on 'that class of portly gentleman known as *Rus in Urbe*'. Many farmers brought their families, attracted by the chance of a few days in town in the weeks before Christmas.

The 1862 Show was the first to be held in the Smithfield Club's brand new Agricultural Hall in Islington and it drew larger crowds than in any year before or since. Judicious breeding had greatly improved the quality of cattle in the sixty-four years since the Club's foundation, and British beef was acknowledged to be the best in the world. An expanding population and increasing prosperity meant that more people were eating more meat than ever before. It was hardly surprising therefore that the Club should be in a position to equip itself with a showplace whose grandeur rivalled that of the larger railway stations. With a glass and iron roof of 150 foot span, the Agricultural Hall was a terminus, with cattle where the trains should have been.

Not everyone was enthusiastic, however. 'It takes some time', commented the *ILN* sourly, 'before the eye can do justice to cattle in their new Christmas home, as the magnitude of the place sadly dwarfs them.' The Hall later won fame as the place where the celebrated American evangelists Moody and Sankey held their revivalist rallies. It is still standing today.

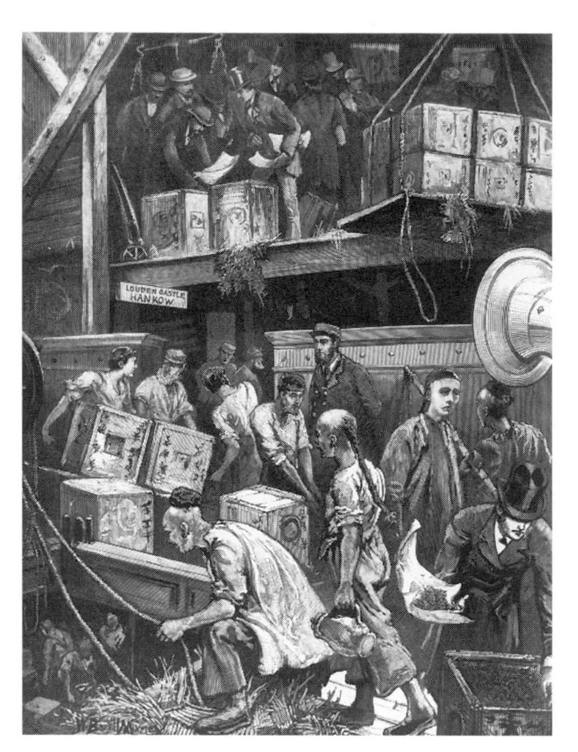

3 Tea ship, London Docks, 1877

An ancient Chinese writer said of tea that 'it tempers the spirit and harmonises the mind, dispels lassitude and relieves fatigue, awakens thought and prevents drowsiness, lightens or refreshes the body and clears the perceptive faculties'. In the early part of the nineteenth century this beneficent drug was brought to London by squat East Indiamen making the journey from Canton in between 180 and 270 days. Later, after the monopoly of the East India Company had been broken in 1834, came faster ships and finally the famous racing clippers, which reduced the time of the voyage to between 90 and 120 days. Their golden age lasted until 1869 when the Suez canal was opened and clippers gave way to swiftly improving steamships as carriers.

The illustration shows the screw steamer *Loudoun Castle* (2,472 tons) in the East India Dock. 'Breaking bulk' is in progress, whereby representative samples of the cargo are made available to waiting dealers.

4 'A parcel of boilers and vats' – Allsopp's brewery, 1888

When Mrs Thrale set out to dispose of her late husband's brewery, Dr Johnson said 'we are not here to sell a parcel of boilers and vats, but the potentiality of growing rich, beyond the dreams of avarice'. Allsopp's brewery at Burton had indeed been a source of great wealth to its owners when in November 1888 the Prince of Wales visited it, escorted by Lord Hindlip (formerly Samuel Allsopp MP), the chairman. It was, however, about to go through a difficult period as a public company during which the quality of its management was to be called into question.

His Royal Highness arrived by rail from Uttoxeter and was welcomed by the 'brewery salute' of a hundred and one fog signals. He visited the malting room and then the mash room, where 'he commenced a brew of pale ale by turning on the water and the malt'. He then inspected the grinding mills, fermenting room, coolers and refrigerators, hop room, racking room and stores. Here he sampled some of the firm's products, including the ale that had been brewed for Captain Sir George Nares and his Arctic expedition fifteen years previously. Lunch followed in the board room.

As the royal visitor was preparing to leave, a group of employees sang 'God bless the Prince of Wales'. We read with a shock that 'the Prince passed out during this interesting manifestation of loyalty', but are relieved when the sentence concludes 'and appeared much impressed by it'.

5 Newcastle upon Tyne, 1877

During the nineteenth century certain towns evolved into cities. Every now and again the *ILN* focused on one such city. In 1877 the one favoured was Newcastle upon Tyne, which had about 3,000 houses in 1800 and 16,000 in 1870.

At Elswick, just off the picture to the left, are the works of Sir W.G. Armstrong & Co., occupying a mile of the river bank, employing about 12,000 men, and helping to arm the world. Prominent on the left is the high-level bridge, by Robert Stephenson, opened in 1849. Its cast-iron arches support railway tracks above and a road below. Trains crossing the bridge curve left into the central railway station, which was opened by the Queen in 1850. To the right of the high-level bridge is the swing bridge (1876), which replaced a stone one of 1781. Armstrong profited doubly from this: he was its contractor, and it allowed him to build at Elswick the largest warships afloat. One of the first ships to pass the bridge had come from Italy to collect a 100-ton gun for the warship *Duilio* (see 102).

In the early nineteenth century the central part of the town was elegantly rebuilt through the cooperation of developer Grainger, architect Dobson, and town clerk Clayton; this prompted Pevsner to describe Newcastle as 'the best designed Victorian town in England'.

6 Iron and steel works near Barrow-in-Furness, 1867

When the railway came to the Furness peninsula in north-west England in the 1840s, William Wordsworth was incensed. He decried the line's directors as 'profane despoilers' for allowing the track to interrupt the view of the ruins of Furness Abbey. But the line had not been built for travellers or sightseers; its purpose was to carry iron ore from inland mines down to the coast, where its terminus was the tiny fishing village of Barrow. Twenty years later Barrow-in-Furness had blossomed into a town of 15,000 people, and had acquired an iron and steel works whose output was at one time greater than any in the land – perhaps even the world.

The Barrow Haematite Company's works was one of the first to produce Bessemer steel on a grand scale: at full blast its eleven furnaces turned out a quarter of a million tons of pig iron a year. Bessemer converters changed half of this into steel, quadrupling its market value. Shareholders had invested one million pounds in the company, and in 1867 they received a dividend of 30 per cent. Town and trade were prospering, and a celebratory dinner was held for 1,200 people at which William Gladstone was guest of honour. 'Some day', he prophesied, 'Barrow will become a Liverpool'.

7 Botallack Mine, 1872

In the mid-Victorian period nearly half the world's supply of tin and copper came from the mines of south-west England. One of them, Botallack on the rugged far west coast of Cornwall, was notable for its diagonal shaft running under the sea to a depth of 250 fathoms (1,500 feet).

The novelist Wilkie Collins (1824–89) described a visit to Botallack in his book *Ramblings Beyond Railways,* published in 1851. Hampered by a miner's suit several sizes too large, he descended by ladder to one of the undersea galleries. A 'strange awe and astonishment' came over the party as they heard and felt the sound of the surf on the rocks 120 feet above, a sound 'so sublimely mournful and still, so ghostly and impressive when listened to in the subterranean recesses of the Earth', that for a time they felt unable to speak in its presence. 'When storms are at their height', Collins was eventually told, 'the roaring heard down here in the mine is so inexpressibly fierce and awful that the boldest men at work are afraid to continue their labour.'

A fall in the price of tin rendered Botallack uneconomic in 1895. It reopened in 1907 but closed again for good in 1914.

8 Mechanical handling, 1873

In the early nineteenth century, coal arriving in the Thames was discharged from sailing colliers into lighters, via weighing machines, by men called 'whippers'. These men were engaged from their best customers by the publicans of some seventy riverside taverns. Whippers worked in gangs of eight plus a foreman. Four toiled in the hold, filling baskets; four worked on deck in two pairs facing one another, each man grasping a rope which passed over a pulley above and was attached to a basket. Acting together, the men climbed four rungs, about a foot apart, of a crude ladder in front of each pair. They lifted the basket a few feet, then, at a word of command, all jumped backwards to the deck. The basket came up with a rush and the foreman seized it and emptied it into the weighing device. If in luck a whipper could get three and a half days work a week and earn 15 shillings. By mid-century his lot had been improved by the creation of a central office for registration and hiring.

In 1860, however, Wm Cory and Son, the biggest coal importers, bought a floating derrick (originally intended to raise sunken ships), fitted it with six hydraulic cranes and moored it in the river, where it discharged coal from steam colliers directly into lighters, without the help of the jumping whippers. Cory acquired another such derrick in 1865. The *ILN* illustration shows one of them, with two colliers, lighters and a tug.

9 Explosion at Ferndale Colliery, Rhondda Valley, 1867

Since systematic recording began in 1850 over 100,000 men and boys have been killed in colliery accidents. The annual number peaked at 1,818 in 1910, then declined and first fell below one hundred in 1970. In the hazardous life of the miner, death or injury could arise from roof falls, moving tubs, shot firing, machinery, shaft accidents, and so on, but it was explosions with great loss of life that focused the attention of the public. Such a disaster occurred at 1.30 p.m. on 8 November 1867 when, of the 328 miners working in the Ferndale Colliery, 178 were killed. The jury attributed this to an accumulation of gas in the mine (for which they blamed the dead manager).

Commenting on a later disaster the *ILN* said, 'the time will come we devoutly trust and confidently believe . . . when those of our countrymen whose lot it is to labour in darkness and in danger deep down below the surface of the earth, and to win from murky depths the mineral which is a principal source of wealth to the country . . . will be able to carry on their toilsome operations, if not as commodiously and pleasantly to themselves, as safely at least as the farm labourers who prepare the soil for our crops, and who reap them when they have arrived at maturity.'

10 Steel pens from Birmingham, 1851

Pen nibs poured from the Birmingham factory of Messrs Hinks, Wells and Co. at the rate of almost a million a day in the 1850s. The factory-made steel nib was still a novelty, swiftly driving the home-made quill pen into extinction. Birmingham pens were shipped 'throughout the civilized world'.

The factory used mass-production methods, with a separate group of workers performing each of the many processes. In the slitting room the only sound was the click of the tool which cut a slit into each nib to make it more flexible and improve the flow of ink. Potted plants on the window-sills suggest that the factory had – for its day – an enlightened management, and this seems to have been the case. The girls were well paid for what must have been numbingly repetitive work: the older ones received between 12 and 14 shillings a week, equivalent to between £24 and £28 in 1993. 'Even at work they are generally well dressed; and to purchase their finery they club together their spare pence, and draw a chance weekly.'

Each girl contributed a penny a week to the hospital fund, enabling her to receive medical treatment when necessary, and a further penny towards the cost of the annual works outing. This was known as a gipsy party, and in 1850 involved an all-day excursion in a fleet of forty-five carriages, festively decorated and accompanied by a 'band of music'.

11 Stephenson's Locomotive Manufactory, 1864

In 1864 the *ILN* published a lengthy panegyric on the history of the railway locomotive in general, and George Stephenson in particular. It claimed that no discovery since the invention of printing had 'exercised so great a change and produced such remarkable and beneficial results to the whole human race'.

Evidence of the extent to which the whole human race was adopting the new form of transport could be found in the order book of the Stephenson Locomotive Manufactory in Newcastle upon Tyne. Among the undertakings for which it supplied locomotives in the 1860s and '70s were the Cape Town Railway, the Great Indian Peninsular Railway, the Victoria and New South Wales Railways, the Great Southern Railway of Buenos Aires and the Imperial Railways of Japan. Stephenson locomotives also hauled trains in Denmark, Holland, Italy, Norway, Egypt, Russia and, from 1882, China.

The works of Robert Stephenson and Co., from which these locomotives came, included the premises in Forth Street where George Stephenson had begun work in 1823. By 1864 the firm was in the hands of a third member of the family, George Robert Stephenson, nephew of the original George. Shown in the engraving is the fitting shop, where highly skilled men assembled locomotives from components made in other parts of the works, completing one 'splendid new locomotive engine and tender of the largest size' every week.

12 The Princess gilds a vase, 1874

The electrolytic deposition of one metal on another was originally observed in 1801, but electroplating as a Birmingham industry can be said to have begun in 1840 when John Wright, a surgeon, and Alexander Parkes, an employee of Messrs Elkington, discovered that a good way to obtain *thick* deposits of *firm* metal of *good colour*, was to pass an electric current through solutions of the *cyanides* of gold or silver. 'The whole trade, both here [Birmingham] and in Sheffield, set their faces against the innovation, and prophesied the ruin of Messrs Elkington . . . but little did they know against what indomitable energy and perseverance they had to contend.' By 1874, when the Prince and Princess of Wales visited Birmingham, electroplating was an established and prosperous component of the city's industry.

13 The Prince and Princess of Wales see the spectacular Bessemer steel process, 1875

The first half of the nineteenth century was the age of wrought iron; the second half was that of steel. The point at which the transition began was marked by a paper read in 1856 by Mr (later Sir) Henry Bessemer (1813–98) to the British Association entitled *The Manufacture of Malleable Iron and Steel without Fuel.* In essence his process consisted of no more than blowing air through molten pig iron, but as an American author wrote in 1890, 'the Bessemer invention takes its rank with the great events that have changed the face of society since the time of the middle ages'.

Such high praise is justified by the fact that Bessemer's mild steel was produced more quickly and more cheaply than wrought iron. It could do almost everything that wrought iron could do, and since it was produced in greater bulk it was particularly suitable for large-scale operations such as shipbuilding and structural engineering. The process had its teething troubles, which were not fully solved for several years, and it was soon supplemented by the Siemens-Martin process; nevertheless Bessemer rightly achieved world fame as a pioneer.

14 At work in a woollen factory, 1883

In the 1880s the manufacture of textiles was Britain's most important industry, and the largest employer of the 'toiling population of our great towns'. Before the industrial revolution, weaving had usually been done by men; now that looms were driven by steam power, they only required 'minders' and these were generally young women. Their job was to watch the loom and stop it every few minutes when the shuttle, which carried the weft (side-to-side) thread, needed replenishing, or if anything else went wrong.

Thirty years had passed since Parliament enacted legislation to limit the work of women in factories to a maximum of ten hours a day, but memories of pale and exhausted mill girls lingered. The *ILN* felt it necessary to reassure its readers that this study of 'the sturdy-looking female hands in a woollen factory' did indeed give 'a truthful representation of their ordinary appearance'.

15 Crossfield's soap works, Warrington, 1886

A Royal Duke is once said to have remarked that it was sweat that kept a man clean. There was indeed a time when toilet soap was not habitually used even by the limited number of people who could afford it. The modest amount that was produced was inevitably taxed, and the kettles in which the fat and alkali were boiled together were locked up at night and excise men held the keys. It is to Gladstone that we owe the repeal of the duty in 1853.

It was the industrial revolution which converted toilet soap from a luxury into an article of common use. It herded people into great cities in which their skins were darkened by soot and ash from the satanic mills, but at the same time soap began to be produced in bulk and at affordable prices.

In the days before the appearance on the scene of W.H. Lever (later Lord Leverhulme, 1851–1925) soap was made by family firms, usually in port areas. One such firm was Crossfield's, established in Warrington in 1814; another was Pears', founded in London in 1769. The latter firm advertised vigorously in the *ILN* and was responsible for the well-known cartoon featuring a tramp saying 'Two years ago I used your soap, since when I have used no other'.

16 Gas holder, London, 1858

The first public company to generate gas centrally for distribution through street mains was the Gas Light and Coke Company, which received its charter in 1812. After a faltering start it eventually got off the ground as a result of the vigour, skill and ingenuity of its chief engineer, Samuel Clegg (1781–1861).

In October 1813 there was an explosion at the GLCC's plant in Peter Street, Westminster. Sensitive to public disquiet, the Home Secretary, Lord Sidmouth, asked the President of the Royal Society, Sir Joseph Banks, to nominate a committee to report on the problem of safety in gas works. When this group (led by Banks himself) visited the works, Clegg, to their no small alarm, knocked a hole in the side of one of the gas holders and applied a light to the issuing gas.

The committee's report was favourable to the company, but it recommended that gas works should be sited at a distance from other buildings or be of small scale. The capacity of gas holders was to be limited to 6,000 cubic feet. The *ILN* picture shows that these recommendations, if they were ever alive, were dead by 1858. The plant is in crowded Bethnal Green and the holder capacity is 2.5 million cubic feet. People living down wind of the works were perhaps consoled by the classical beauty of the gas holder – note its Corinthian and Tuscan columns.

17 Electric light at the British Museum, 1890

In 1889, following pressure in Parliament for the evening opening of the British Museum, £5,000 was voted for the provision of the necessary electric lighting. The resulting system, designed by W.H. Preece FRS, consisted of two independent generating sets, active and stand-by, each consisting of a steam engine driving a pair of Siemens dynamos in parallel, capable of supplying 450 amperes at 130 volts. Both arc and incandescent lamps were used, depending on the nature of the gallery to be lit. The installation was demonstrated by the trustees in January 1890 at a private viewing by persons prominent in science, art, literature, law and religion.

Evening attendance figures were disappointing. Throughout February 1890 the average was 635, in March 367, in November 145. Sometimes the number of attendants present exceeded the number of visitors. 'The light shines alone on the placid mummy and the still more placid policeman,' wrote one reporter and he added that 'the long galleries are as silent as any passage in a rock temple'. Part of the trouble was due to the fact that the Museum closed at 5 p.m. and then reopened from 8 p.m. to 10 p.m. for the evening session. Furthermore, evening visitors would find either the eastern or the western galleries open but not both, so that 'one may go there with the intention of studying early English sundials and have to put up with mummies'. Evening opening ended in 1898.

18 Victoria Embankment as forecast, 1867

Several proposals for embanking the Thames were made in the nineteenth century. Colonel (later General Sir) Frederick Trench (1775–1859) started to form a company to build an embankment, but it failed to get Parliamentary approval. The artist John Martin (1789–1854) proposed not only to embank the Thames but also to pass through the reclaimed land a main sewer that would divert minor ones which were discharging into the river. He gained moral support but little more from a committee that included forty-four MPs, eighteen FRSs (including Faraday and Wheatstone) and six RAs.

The later plan of Mr (later Sir) J.W. Bazalgette (1819–91), Chief Engineer of the Metropolitan Board of Works, had virtues that were obvious to all, since an embankment between Westminster and Blackfriars would greatly facilitate his own approved scheme for the drainage of London. This involved a low-level sewer hugging the Thames, which would have meant digging up the Strand and Fleet Street. But, given an embankment, the sewer could (as in Martin's scheme) pass through the reclaimed land; so too could a tunnel to carry the proposed Metropolitan District Railway, as well as a subway to carry services such as water mains, gas pipes and telegraph cables, and these could be maintained without digging holes in the road to get at them. The well-known *ILN* picture that illustrates these advantages shows also a tunnel for an atmospheric railway linking Charing Cross and Waterloo, but nothing came of this.

1
2
4

19 The Vyrnwy Dam in 1888

In 1864 a dam was completed at Dale Dyke in Yorkshire, forming a reservoir that would hold water for the city of Sheffield. The dam was an earth one, and no sooner had the reservoir filled than the dam began to subside, releasing 200 million gallons of water into the valley below and causing the deaths of 250 people.

It is thus no surprise that when the engineers of the Liverpool Waterworks were charged with damming the Vyrnwy valley in North Wales they opted for a masonry dam, firmly anchored to the rock beneath. Stone blocks up to ten tons in weight were set in a bed of Portland cement to form a solid structure 161 feet high and nearly a quarter of a mile in length. The engraving shows the outer face as the dam neared completion in the autumn of 1888.

The valves were closed on 28 November and ten days later the new Lake Vyrnwy had reached a length of four miles. The village of Llanwyddin was submerged, but Liverpool Corporation gave 'full compensation' to the forty or fifty families displaced, erected a new parish church above the waterline, and transferred to it the inhabitants of the graveyard. Seventy-seven miles of aqueduct completed 'one of the grandest engineering works of our time' ensuring for the people of Liverpool 'an abundance of the purest water'.

20 Drinking fountain, 1862

Filthy river; filthy river
Foul from London to the Nore.

Thus *Punch* in 1858 addressed Father Thames, that giant sewer from which for centuries many Londoners had derived their drinking water. Thanks to legislation of 1852 requiring that such water be extracted above Teddington and filtered, it was no longer as lethal as it had been, but for the poor it was still dispensed fitfully and reluctantly from sparsely provided standpipes, and therefore many continued to opt for beer and gin, particularly when the alehouse was usually the only affordable source of refreshment.

So it was both as a sanitary and a temperance reformer that Samuel Gurney MP (1816–82) established the Metropolitan Free Drinking Fountain Association in 1858. Fountains supplying pure cold water, variously financed, sprang up all over the city, some erected as memorials to the great and the good. Perhaps the grandest was that established in Victoria Park by the philanthropist Angela Burdett Coutts, at a cost of £6,000. Artistically it was a noble piece of eclecticism, combining Venetian, Moorish and Renaissance elements.

21 Abbey Mills pumping station, 1868

Originally, London's human sewage either accumulated in cesspits, was cast into the open street, or passed by way of its many tributaries (Effra, Falcon, Fleet, Tyburn etc) into the Thames. By the nineteenth century, brick sewers were being constructed. At first they were to carry rainwater into the river (since heavy rain usually produced flooding somewhere in the city) and citizens were prosecuted for using them for offensive matter; later, in order to get rid of cesspits, citizens were prosecuted for not so using them. Hence the Thames became grossly polluted; it stank, and by the mid-nineteenth century it was clear that something had to be done.

The great man who did it was J.W. Bazalgette, Chief Engineer of the Metropolitan Board of Works. In a period of seven years and at a cost of £4.6 million he constructed five brick sewers running approximately east-west, intercepting the scores of existing sewers and carrying their contents to the Thames below London. North of the river there were three, uniting into the Northern Outfall to discharge at Barking Creek. South of the river there were two, discharging through the Southern Outfall at Crossness. Where necessary, pumping stations were erected whereby the sewage was lifted so as to run by gravity at a speed adequate to keep the channels flushed. Some astonishing architecture characterized these stations; that at Abbey Mills, Stratford, is illustrated.

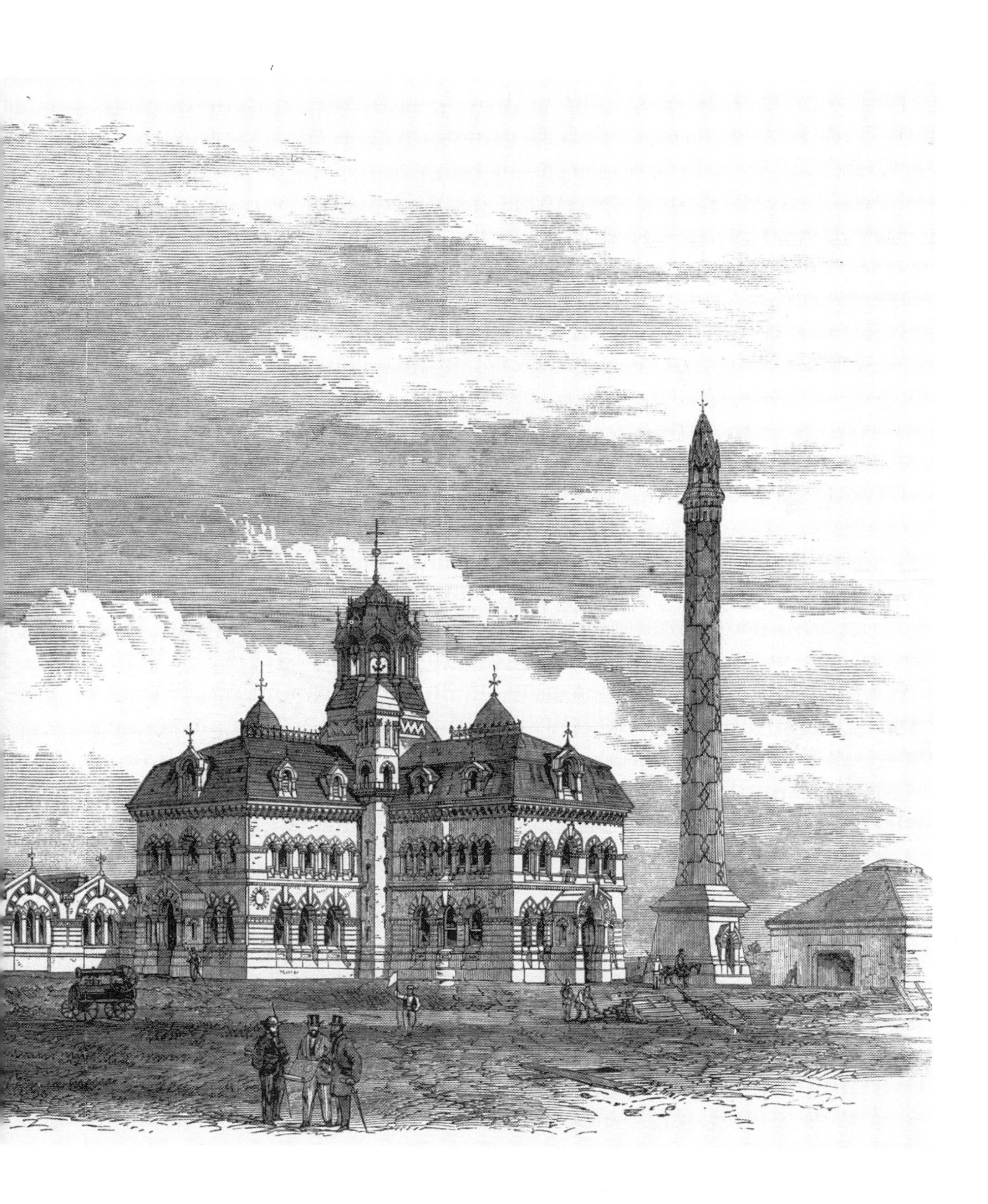

22 Metropolitan Firc Brigade at practice, 1868

The commonest type of fire is extinguished by cooling the burning material and denying it oxygen. Water does this admirably, but for centuries the only technique by which it could be applied was by means of buckets, perhaps passed along a human chain. In the early seventeenth century the pump became available and when it was mounted on a sled or (later) wheels, it became a fire engine. In mid-century came the pressure dome whereby the pumped water created air pressure which then propelled it as a smooth jet towards the fire. But the chain of buckets was still needed to fill the engine's cistern. The hose (*c.* 1690) was the answer, and while one hose allowed the pump to draw water from a remote source another enabled the firefighter to aim his jet precisely.

The mature fire engine developed from these elements was brought into action by men or horses. But only men could work the pump and they did so as a team of (say) twelve, with six on each side working an arch lever. Then came the steam engine, tried experimentally in 1829 but exploited fully only in the 1860s. However, in 1870, two years after our *ILN* picture, Captain (later Sir) Eyre Massey Shaw, commander of the Metropolitan Fire Brigade, had at his disposal twenty-five steam and eighty-five manual fire engines. So manual ones were by no means obsolete, one reason for favouring them being that the powerful steam pumps sometimes destroyed with water the objects they were trying to preserve from fire.

METROPOLITAN

23 The lady with the lamp, 1855

Florence Nightingale (1820–1910) was born into an upper-class family and might have been expected, after a narrow education and the acquisition of some accomplishments, to attract high bids in the marriage market, consistent with her external beauty and the wealth of her parents. Florence would have none of this; against parental opposition she sought self-fulfilment through some form of successful service, perhaps nursing. The opportunity came in October 1854, seven months after the outbreak of the Crimean War, when she was invited to become 'superintendent of the female nurses of the hospitals in the east'. She left for the barrack hospital at Scutari (Constantinople), one week after her appointment, with thirty-eight nurses.

Those at home learnt of her as compassionate and caring from the story of her visits to the wards at night with her little Turkish lamp, from the illustration of one of these visits in the *ILN*, and from the soldier who wrote 'we lay there by hundreds, but we could kiss her shadow as it fell'. Many, perhaps most, know no more about her than this, and do not realize that her victory against the stultifying medical orthodoxy of the time, her conversion of a huge stinking and verminous pest-house into a tolerable hospital, and actions such as her procurement without authority of clothes for patients admitted in rags, could be achieved (when tact had failed) only by the exercise of combative and steely elements of her character. These make her more admirable, if less lovable, than the lady with the lamp.

24 'There is death in the cup', 1892

The *ILN's* roving artist in Europe, Mr J. Schönberg, sent a sketch back from Hamburg in 1892 from which W.H. Overend constructed a picture whose title was guaranteed to seize the reader's attention. Schönberg had 'witnessed, in a shop where he called to make some purchase, a little girl drinking freely of water, the purity of which might be doubted. A few hours later he was told that this child was dead.' Cholera had claimed another victim.

The outbreak was a serious one. In Britain, ships arriving from Hamburg and other affected ports were intercepted and cholera cases immediately isolated. As a result the 1892 cholera epidemic never came here, nor has there been one since.

25 Robert Koch and the 'cure for consumption', 1890

Robert Koch (1843–1910) laid the foundations of the science of bacteriology, and discovered the microscopic agents which cause anthrax, tuberculosis, and cholera.

His portrait appeared in the *ILN* in 1890 in unfortunate circumstances. Working in his Berlin laboratory to find a cure for consumption (known also as tuberculosis), Koch fell into the trap of announcing positive results too soon. The excitement was immediate and frenzied. Within days 1,500 doctors had converged on Berlin from all over Europe to find out more. But it was not the cure, and at a time when consumption still claimed more than a thousand British lives every week, the subsequent disappointment was intense. Koch's reputation was only temporarily dented, however, and he went on to receive a Nobel Prize for his work on tuberculosis.

26 Louis Pasteur, 1884

During the course of a working life of forty-eight years, beginning with important studies in crystallography, Pasteur (1822–95) discovered the agent of lactic acid fermentation (as in the souring of milk); elucidated the part played by yeast (a living organism) in alcoholic fermentation; showed that there was no evidence for the spontaneous generation of minute living organisms, those which had been observed having come from the air; discovered the nature of the ferment that produces vinegar; identified the ferments that lead to the diseases of wine, and showed that these diseases could be prevented by heat treatment; studied the diseases of silkworms and suggested remedial measures; studied industrial brewing (with a view to making France more competitive in this field) and suggested improved procedures; identified the bacterium responsible for anthrax and showed that the disease could be treated by vaccination; made similar studies of chicken cholera; proved that rabies was a virus disease, and had success in treating human beings.

Towards the end of his life he said to his grandson, 'Ah, my boy, I wish I had a new life before me! With what joy I should like to take up again my work on crystals.' Might he have become the Galileo or Newton of his field had he stayed with pure science?

27 Lord Lister, 1897

➸

Joseph Lister (1827–1912) was the second son of Joseph Jackson Lister, Quaker, wine merchant, FRS and inventor of the achromatic microscope objective. Barred by his faith from the older universities, he received first a general and then a medical education at University College London, leading to election as FRCS in 1852. After experience of surgery in Edinburgh he became in 1860 Regius Professor of Surgery at Glasgow University, in 1869 Professor of Clinical Surgery at Edinburgh University and in 1877 Professor of Surgery at King's College London. He was created baronet in 1883 and baron in 1897, and was President of the Royal Society from 1895 to 1900.

It was at a Royal Society dinner in 1902 that the American ambassador, addressing Lister, said, 'My Lord, it is not a profession, it is not a nation, it is humanity itself which, with uncovered head, salutes you.' The work of Lister which called forth this wholly deserved panegyric was his introduction first of antiseptic, then of aseptic surgery. Recognizing that infection of a wound might be attributed to airborne micro-organisms of the kind discovered by Pasteur, he devised procedures, originally based on the use of carbolic acid, which minimized the number of such organisms that reached the wound and destroyed those that did. The procedures of asepsis have changed; the principle remains and humanity owes this to Lister.

CO_2
CO_2
CO_2

The Queen opens the Manchester Ship Canal, 1894

Communications and Transport

The electric telegraph was 'the wonder of the age' when the *ILN* first appeared. It grew alongside that other wonder, the railway, but did not stop at the coast. The struggle to lay a usable Atlantic cable provided pictures and a story scarcely rivalled elsewhere in the history of technology. Our picture of the telephone, by contrast, shows it as simply a plaything.

Water transport was represented in the *ILN* by numerous views of launchings. We chose to concentrate on the birth and subsequent adventures of just two great ships, *Great Britain* and *Great Eastern*, both associated with the name of Isambard Kingdom Brunel.

Railways dominate the age. In the earlier years we see the steam locomotive cheekily challenged by other means of propulsion – air pressure, or horses on a treadmill. Later it is the grander images which impress, of termini, a grand hotel, and a Pullman dining-car. We also glimpse the spread of railways in Asia and North America. Bridges are prominent in a wide variety of design, including one which collapsed.

Representing road transport, horse-drawn vehicles crowd over and under London's Holborn Viaduct, and the steam carriage and electric tram put in an appearance; the bicycle provides fun at several levels in society. Only as the century ends does the motor car come light-heartedly into view. Our chosen picture, portentous in itself, also signals an impending revolution in the technology of graphic communication: the displacement of the craft of wood engraving by techniques of reproduction requiring less labour and perhaps less skill.

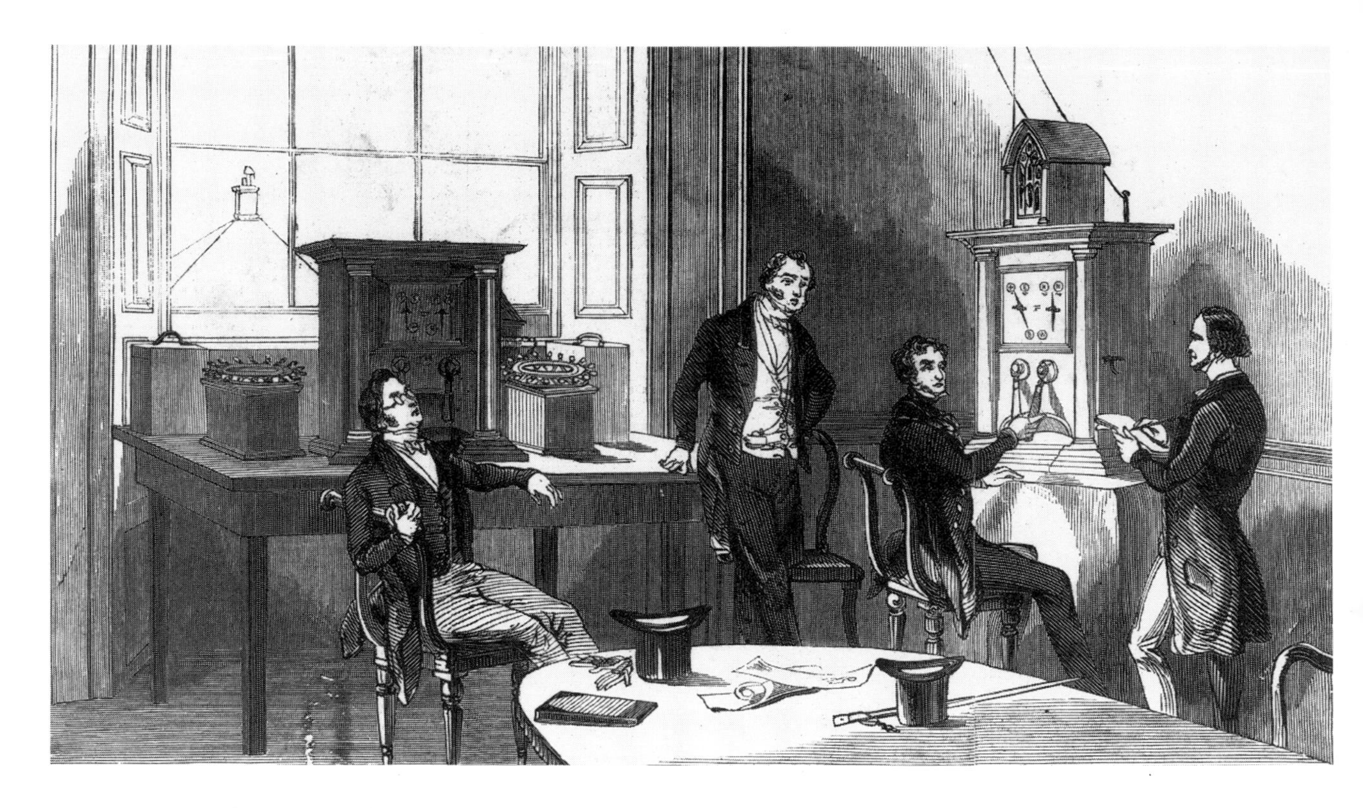

28 Chess via the electric telegraph, 1845

'To confer at the distance of the Indies by sympathetic conveyances may be as usual to future times as to us in literary correspondences.' Thus in 1665 the telegraph was foreseen.

The 'sympathetic conveyances' needed were a generator of signals, a medium of transmission, and a detector. Volta provided the first when in 1800 he invented the electric cell; Oersted supplied the third when in 1820 he showed that a suspended magnet could be deflected by an electic current, and the second had been supplied by eighteenth century 'electricians' when they distinguished experimentally between conductors and insulators. For practical telegraphy there were also human requirements. In this country the necessary vision, faith and entrepreneurial initiative were supplied by William Fothergill Cooke (1806–79), and the equally necessary scientific knowledge and practical ingenuity by Charles Wheatstone (1802–75).

In urgent need of the invention were the new railways under construction, but the first to recognize this was the Great Western, when the Paddington–West Drayton telegraph link opened in 1839. For the rope-hauled Blackwall Railway (1840) telegraphy was essential. In 1845 came the completion of the 90-mile link between Nine Elms and Gosport on the London and South Western Railway, giving the Admiralty two of the four circuits. The *ILN* illustration shows a chess doubles match between Portsmouth and London being played over this link. One of the Portsmouth players was Mr Staunton, chess editor of the *ILN*.

29 The time ball in the Strand, 1852

From 1833 the chronometer of a ship in the Thames could be checked by observing the fall of a time ball from the top of a pole on the eastern turret of the Royal Observatory, Greenwich. From 1836 a weekly time service of a more personal type was supplied to the chronometer makers of London by Mr J.H. Belville, the son of a French lady whose husband had been guillotined in the revolution. Checking his Arnold chronometer at Greenwich, where he was an assistant, he carried it around his circle of subscribers. When he died in 1856 the service was continued by his wife Mrs Belville, and when she retired in 1892 it was taken over by her daughter Miss Ruth Belville who carried on until the 1930s. According to de Carle her business with a client began thus: 'Good morning, Miss Belville, how's Arnold today?', to which she might reply 'Good morning! Arnold's four seconds fast today'. It is remarkable that this service should survive the introduction in 1852 by G.B. (later Sir George) Airy, the Astronomer Royal (see 81), of a national electrical time service. This was provided by means of an electric master clock, checked daily by stellar observations, which sent electrical impulses every second to operate local slave clocks, and hourly impulses via the railway telegraph lines to the Central Telegraph Station of the Electric Telegraph Company for further distribution, locally to a time ball in the Strand, and throughout the country via the public and railway telegraph services.

30 Atlantic cable – failure, 1857

Every aspect of the great enterprise of laying the Atlantic cable makes a good story. It was a remarkable example of international capitalistic enterprise; it involved constructive government co-operation; mechanical engineering was called upon to meet entirely novel demands; electrical engineering, only just born, faced practical problems whose theory had yet to be worked out; failure after failure after failure was finally crowned with success.

For the first attempt in 1857 the conducting core of the cable was made by the Gutta Percha Company in London. Half of it was armoured by Messrs Glass, Elliot and Company at Greenwich, the other half by Newall and Company at Birkenhead. From America came USS *Niagara,* 5,200 tons, to take on the Birkenhead cable. For the Greenwich cable the Royal Navy lent the screw line-of-battle ship HMS *Agamemnon,* 3,200 tons, which had been launched at the Royal Dockyard, Woolwich, in 1852, and which in 1854 had been the flagship of Sir Edmund Lyons at the bombardment of Sebastopol. The illustration shows cable being loaded into the hold of *Agamemnon.* The monstrous top hat seems somehow symbolic.

The cable laying started at Valentia, Ireland on 6 August 1857, from the hold of *Niagara.* After three-quarters of an hour the cable parted and a fresh start had to be made. After 334 sea miles of cable had been paid out, it parted again after an incorrect application of the brake mechanism, finally bringing the 1857 attempt to an end.

31 Atlantic cable – success then failure, 1858

In 1858 *Niagara* and *Agamemnon* made a second attempt, equipped with improved paying-out machinery, a better method of storing cable, and the sensitive mirror galvanometer of Mr (later Sir) William Thomson as detector. In mid-Atlantic on 26 June 1858, *Niagara* started laying towards Newfoundland and *Agamemnon* towards Ireland. Breaks of cable necessitating restarting occurred at separations of 6 miles and 80 miles, and after another break at 255 miles the ships returned to Queenstown to restock with food and fuel.

A final effort began on 29 July 1858. This time success was achieved: Europe and America were connected electrically. There was much rejoicing, not perceptibly dimmed by the fact that Queen Victoria's ninety-eight word message to President Buchanan on 13 August took sixteen and a half hours to transmit. After four weeks, however, the cable failed.

After an inquest, a government-ordered study of the problem of submarine telegraphy by an expert committee, a fallow period, and a restructuring and refinancing of the promoting company (still under its indefatigable leader Cyrus W. Field), a further attempt was planned for 1865. This time the whole of the cable was to be carried on one ship, namely Brunel's *Great Eastern,* here seen lying in the Medway and taking in cable brought from Greenwich in a hulk.

32 Atlantic cable – failure, 1865; double success, 1866

Accounts of how the *Great Eastern* entered the story differ in detail, but it would appear that the vessel was acquired by a syndicate, headed by Mr (later Sir) Daniel Gooch MP, who had been the locomotive superintendent of the Great Western Railway. He offered to charter it to the sponsoring company of the Atlantic cable for nothing if the attempt failed or £50,000 of company stock if it succeeded.

The *Great Eastern*, with two attendant warships, started from Valentia for Newfoundland on 23 July 1865. There were two scares when continuity failed, in each case because a metal pin had penetrated the cable. Finally, after about 1,200 miles had been paid out, a break occurred and the cable was lost. Attempts to grapple showed promise but ultimately failed.

After more finance had been raised and new cable taken on board, the *Great Eastern* left Valentia once again on 13 July 1866 and arrived at Heart's Content, Trinity Bay, Newfoundland on 27 July. Then the ship returned to the place, indicated by buoys, where the 1865 cable had been lost, and successfully recovered its end. Then occurred perhaps the most dramatic moment of the whole enterprise. Was this cable sound? Was there still continuity with Valentia? The illustration is said to show the tension among the spectators as the test was made. There was continuity; the recovered cable was extended to Newfoundland and the Anglo-American Telegraph Co. now had two Atlantic cables.

33 New Year greetings by telephone, 1882

Over the years a number of inventions have been described as 'a solution in search of a problem'. The telephone, surprisingly, is one; many people were unable to foresee how it might be used. Sir William Preece, for example, engineer-in-chief to the Post Office, spoke in sceptical terms of the usefulness of the new device in a country which had 'a superabundance of messengers, errand boys and things of that kind'. The instrument in his office was 'more for show', he said; 'If I want to send a message, I . . . employ a boy to take it.'

One proposed application was the telephonic seance, shown here. The 'miraculous hearing tubes' are being used to bring the words of a distant speaker simultaneously to the ears of everyone in the room (apart from the elegant couple on the right, who seem to have other ideas).

The *ILN* report suggests that such a performance of applied physical science might be particularly appropriate for the exchange of New Year greetings, adding with heavy-handed whimsy that difficulties would arise if the telephone ever achieved international range; since the clock does not strike twelve simultaneously all around the world, the instantaneous exchange greetings 'may prove to be not a very simple matter, let Science do whatever she will'.

34 SS *Great Britain* floated out, 1843

Isambard Kingdom Brunel's *Great Western* (236 feet, 2,300 tons), which was launched at Bristol in 1837, proved in the following year that a wooden paddle steamship of sufficient size could cross the Atlantic and have fuel to spare. Brunel soon began to plan a bigger steamship on the same lines, but experience gained during the visits to Bristol, firstly of a small iron vessel (*Rainbow*, 1839) and secondly of a pioneer screw-driven ship (*Archimedes,* 1840), led to a change of mind and the big ship that finally emerged, *Great Britain* (320 feet, 3,500 tons), was of iron and propelled by a screw and sails, the latter on six masts.

Those most concerned with the creation of the ship were Brunel, its designer; Captain Claxton, ex-naval officer, Brunel's right arm; William Patterson, ship builder; and Thomas Guppy of the Great Western Railway Steamship Company which built, owned and operated the ship.

Great Britain was built in a dry dock off Bristol's 'floating harbour'. Floating out, on 19 July 1843, in the presence of Prince Albert, was rather less exciting than a slipway launch: the dock was filled and the hull was towed across to the Gasworks Wharf for installation of the engines. There was considerably more drama when the ship, of beam 51 feet, had to pass through a lock only 44 feet wide into the Cumberland Basin, and again when she finally passed into the Avon on 12 December 1844.

35 SS *Great Britain* aground, 1846

Great Britain left Liverpool at 11 a.m. on 22 September 1846 for her fifth voyage to New York. At about 9.30 p.m. she ran aground in Dundrum Bay, Ireland. Captain Hosken prevented panic; all got off, unharmed, with their luggage, next morning. In the illustration the people on the beach are engaged in lightening the ship, preparatory to an attempt to float her off.

After wintering behind a breakwater designed by Brunel, consisting primarily of brushwood, *Great Britain* was refloated and towed to Liverpool, where she was sold. Rebuilt with two funnels and four masts she began in 1853 a career that included thirty-two voyages to Australia and four to New York, interrupted by trooping for the Crimean War, and again for the Indian Mutiny.

In 1876 she was laid up in Birkenhead and in 1882 she was bought by a company which converted her into a cargo ship with sail propulsion only. In this form she made two voyages to San Francisco. On a final voyage (1886), following bad weather at Cape Horn she ended up at the Falkland Islands. There she stayed for fifty years as a hulk for wool and coal before finally being scuttled. But that was by no means the end of the story; as readers know she is now in the dock at Bristol where she was built. The driving force behind this happy return was supplied by Dr Ewan Corlett, in whose book *The Iron Ship* the fascinating story is told with magisterial authority.

36 The *Great Eastern* – attempted launch, 1857

I.K. Brunel (1806–59) was forty-six years old when he conceived the idea of building a steamship that would be enormously larger than any then afloat. The scale of the proposed increase was not megalomaniac; the proposal could be backed by sound commercial reasoning since the full advantage of steam over sail for very long voyages could be achieved only if the steamship could carry enough coal for the whole journey.

The idea was approved by the directors of the Eastern Steam Navigation Company. Brunel was appointed engineer, John Scott Russell was awarded a contract for the hull and paddle-wheel engines, and James Watt and Co. that for the screw engines. At Scott Russell's Millwall yard the ship had to be built on the bank of the Thames for sideways launching. Difficulties of construction, aggravated by tension between engineer and constructor, eventually led to the bankruptcy of Scott Russell.

Launching was meant to occur on 3 November 1857 with the ship, supported on two massive wooden cradles, being pulled down two runways into the river, the motion being checked by chains wound round massive drums on the landward side (prominent in the illustration). In fact the launch failed. Later attempts, with hydraulic rams of ever increasing power, moved the ship only short distances and it was not until 31 January 1858 that the hull of the *Great Eastern* floated.

GREAT EASTERN
GREAT WEST
GREAT BRITA
SALTASH.
BRUNEL

37 The Death of Mr Brunel, 1859

News of the explosion on the *Great Eastern* was brought to Brunel in London. Of such mishaps he had earlier written that 'the most valuable experience is that derived from failures and not from successes', but this time he was too ill to react either to success or failure. Two days later he died, in his fifty-fourth year, possibly of the kidney complaint now known as Bright's disease. He was given a private funeral, attended by a number of 'gentlemen distinguished in engineering science', and buried at Kensal Green cemetery.

Among the laurels which frame his portrait the *ILN* artist has placed crude miniatures of five of Brunel's greatest works. Three of them are pictured elsewhere in this book (34, 36, 48); the others are the *Great Western*, the first steam-powered ship to ply the Atlantic (top left), and the Hungerford suspension bridge over the Thames (bottom right), whose chains were later reused at Clifton (63).

Brunel's father, Sir Marc, had been born in France, and Brunel himself studied mathematics and science in that country. A twentieth-century engineer has suggested that his success derived in part from his inheritance of two complementary traditions, 'the practical pragmatic emphasis in Britain and the high value which France gave to intellectual accomplishment' (Crook, 1990). To his French admirers, Isambard Kingdom Brunel was quite simply 'the Napoleon of engineers'.

38 The *Great Eastern* – departure for New York, 1860

While the *Great Eastern* was being fitted out at Deptford it was decided that she would be moved to Weymouth and from there a trial trip would be made into the Atlantic. She left on 7 September 1859. Two days later, off Hastings, 'there was, throughout the whole vessel a sound of most awful import, quickly followed by the fiendish hissing of disported steam which, freed from the confines of its iron cage, and breathing death, mocked agony to scorn'. A steam explosion had occurred blowing one of the funnels 30 feet into the air and killing five stokers.

After repairs at Weymouth, a trial run to Holyhead, and wintering at Southampton, she began her maiden voyage to New York on 17 June 1860. She had been designed to carry three thousand passengers, but so far had been fitted up for 300. There was however, no overcrowding since the number of paying passengers was thirty-eight. According to Dugan the cargo consisted of 500 gross of London Club Sauce. The illustration shows her leaving Southampton Water.

Thus the *Great Eastern* started on the luxury transatlantic run for which she was not designed and for which she was not suited. She was never a commercial success, but she was one of a Brunel trio of which Caldwell says 'each of his three great ships . . . advanced in some major and lasting way the science and practice of naval architecture'. (For her work as a cable ship see 31 and 32).

39 Beachy Head lighthouse, 1884

The earliest post-Roman lighthouses in Britain, e.g. Dungeness (1615), used burning coal as illuminant. This had its disadvantages: coal had to be got to the top of the tower and ash to the bottom, over-enthusiastic use of the bellows burnt the brazier, and a ship might be endangered in daytime if the smoke rising from a lime kiln was mistaken for that from a lighthouse. Hence within the lantern of Smeaton's great feat of engineering, the fourth Eddystone lighthouse of 1759, there burned only a ring of twenty-four candles, giving a visibility of less than five miles. Primitive oil lamps were available but they blackened the lantern and were shunned.

Blackening was caused by incomplete combustion, which was almost completely eliminated in 1784 when the Swiss Ami Argand invented a lamp with a cylindrical wick up which flowed a current of air. Argand may have also invented the glass chimney, but in France this is associated with the name of Quinquet.

The Beachy Head lighthouse, shown here, was built on Belle Tout Cliff between 1828 and 1831. Illumination was by a battery of Argand lamps, each backed by a parabolic mirror of copper plated with silver, the whole rotating once in two minutes. Visibility was 22 miles. In the picture one keeper cleans a mirror while the other is at the winding handle of the clockwork motor. The lighthouse was subject to blanketing by fog and was replaced in 1902 by the present one at the foot of the cliff.

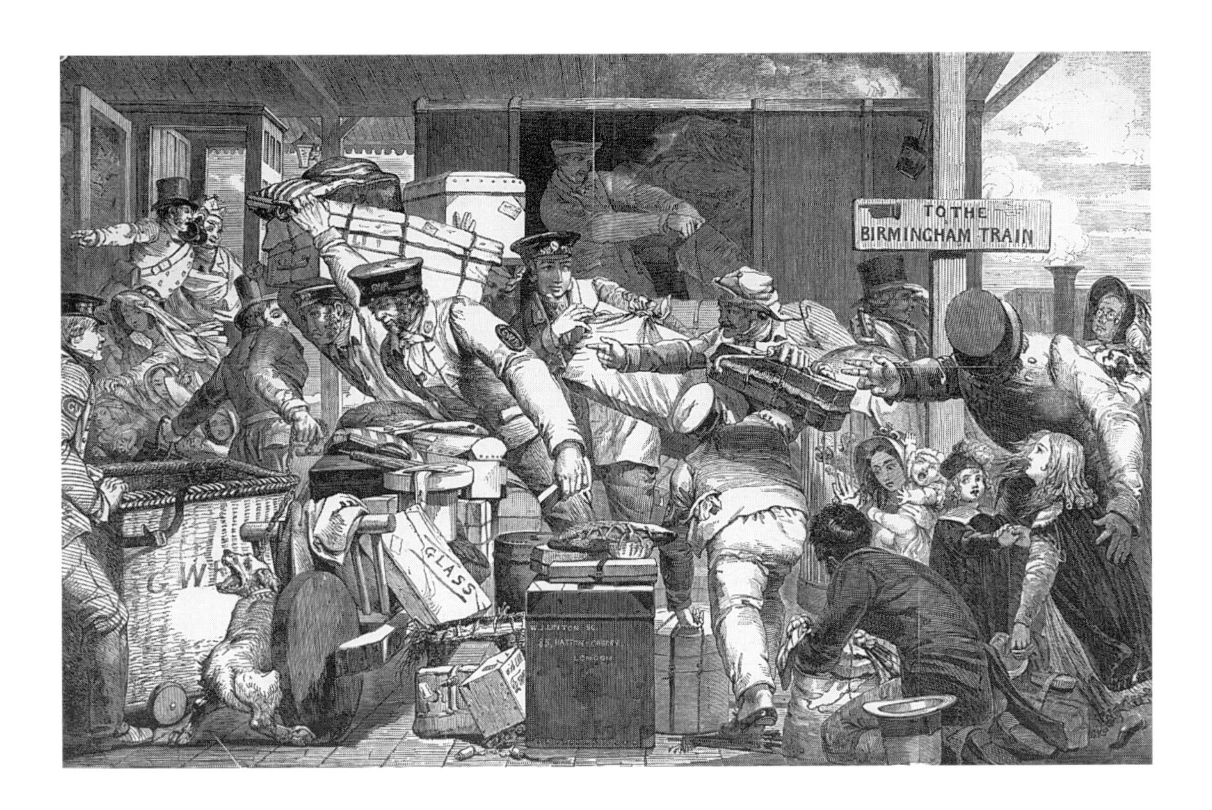

40 The break of gauge at Gloucester, 1846

When George Stephenson built the world's first public steam railway in 1825 he set the rails 4 feet 8 inches apart, retaining a separation that had evolved for earlier horse-drawn wagonways. Modified slightly to 4 feet 8½ inches it became known as standard gauge and was widely adopted. Brunel defied the convention, however, by constructing his Great Western Railway to a gauge of 7 feet ¼ inch, arguing that this would give passengers a swifter, safer, and smoother ride. So began the 'Battle of the Gauges', fought first at Gloucester, where the meeting of different gauges meant that passengers between Birmingham and Bristol had to change trains.

The resulting inconvenience and confusion were lampooned by the *ILN* in its issue of 6 June 1846. Readers were invited to identify with the gentleman whose dog is about to have a paw crushed, while his wife's medicine chest spills open onto the platform. A foreigner (signified by outlandish headgear) swears at the porter who has carried off his luggage, and in the background a railway policeman bawls into the ear trumpet of a deaf and uncomprehending traveller. Livestock, we are told, presents a particular problem, and the gentleman resolves never again to take his horses by rail where there is break of gauge.

Brunel's broad gauge lingered until May 1892 when the last 65 miles of track were relaid to standard gauge over a single weekend by a team of 4,700 navvies.

41 Proposed atmospheric railway station, 1845

The Epsom extension of the London and Croydon Railway was planned to be 'atmospheric' in more senses than one. Architecturally, an aura of antiquity would be engendered by stations built 'in the style of the half-timbered manor houses of the middle ages', and crowned with chimneys resembling 'the bell towers of the early Gothic churches'. Technically, the trains would be propelled by atmospheric pressure.

Traffic on the London and Croydon line had initially been hauled by locomotives, but the directors soon decided to convert the line to the atmospheric system, having seen it successfully applied to a railway in Ireland. An iron pipe, visible in the illustration, was laid between the rails. A piston attached to the train ran inside it, access being gained through a slit along the top of the pipe, sealed with a system of leather flaps. A stationary steam engine, disguised beneath the Gothic bell tower at every station, evacuated the tube ahead of the train, leaving atmospheric pressure behind the piston to drive it forward.

In October 1845 the first ten-carriage atmospheric train reached a speed of 52 m.p.h. and it seemed that the steam locomotive might be facing real competition. However, in regular service the system proved unreliable, principally because of difficulty in keeping the pipe airtight, and the directors soon abandoned it. When the extension to Epsom opened in May 1847, just in time for Derby Day, trains were again hauled by steam locomotives.

42 Pneumatic railway at the Crystal Palace, 1864

In 1810 one George Medhurst proposed a mode of transport in which a carriage was to be blown along rails in an iron tube by the action of compressed air. The pneumatic railways that were actually built later in the century used a different principle (see 41), but Medhurst's idea was revived in 1860 by the Pneumatic Despatch Company for the transport of small goods, in particular mail.

T.W. Rammell, the company's engineer, started with a successful experiment on a riverside track at Battersea. Subsequently a tunnel for mail, about 4 feet in diameter, was constructed between Euston and a London district post office, and ultimately Euston was connected to the General Post Office by this means, but without lasting success.

Meanwhile Rammell demonstrated the feasibility of the method for public transport on a 600 yard track at the Crystal Palace, Sydenham. The carriage, with some thirty passengers, descended a slope into the tunnel, whose doors closed behind it, and then air, delivered by a huge fan, blew it to the far end. To effect a return by the agency of atmospheric pressure the fan was put into reverse and exhausted the tunnel. The carriage carried a necklace of bristles to limit transmission of air past it.

43 Tubular bridge – rehearsal, 1848 . . .

When Robert Stephenson was appointed Engineer-in-Chief for the Chester and Holyhead Railway in 1845, he was confronted with the problem of how to bridge the Menai Straits, separating the mainland of North Wales from the Isle of Anglesey. After extensive experiment, and many nights when he 'would lie tossing about seeking sleep in vain', Stephenson proposed a design which was as simple as it was audacious. A single tubular girder, constructed from wrought-iron plates, would carry each track across the Straits, and the trains would run inside.

A few miles before the railway reached the Menai Straits it had to cross the narrower Conway estuary. This gave Stephenson the opportunity to try out his new design on a smaller scale. For the Conway bridge, two 1,300-ton metal tubes were constructed nearby, and on 6 March 1848 the first, shown here, was floated on pontoons out to its destination between piers already built on either bank, where it would eventually be raised to the correct height using hydraulic rams.

Among those standing atop the tube with Robert Stephenson as it was manoeuvred into position were his father George, and I.K. Brunel, who had come to lend moral support. Directing the exercise through a megaphone was Captain Claxton. On its successful completion workmen and spectators gave three hearty cheers, and Claxton smashed his 'speaking-trumpet' and pitched it into the Straits.

44 . . . and grand performance, 1850

A year after his successful bridging of the Conway, Stephenson began the much more arduous and hazardous operation at Menai. Here each tube was to be 1,513 feet in length, made up from shorter lengths, and had to be raised 100 feet above the water. The two greatest spans would each be seven times the length of any girder bridge built before.

The task was completed in March 1850 when Stephenson himself hammered home the two-millionth – and last – rivet, accompanied by 'the waving of hats and the deafening acclamations of the work-people'. The government inspector ran a 240-ton train with three locomotives across the finished bridge and concluded that 'all parts of the stupendous work obeyed the calculated requirements'.

The engraving shows the Britannia Railway Bridge from the Anglesey shore, with Telford's earlier road bridge beyond. The two structures became lasting monuments to engineers who were heroes of their day. When Stephenson died, nine years after completing his greatest work, the Queen 'expressed her sympathy with the loss the nation had sustained'. The public lined the entire route of his funeral cortège and an immense crowd assembled at Westminster Abbey, where he was laid to rest alongside Telford in the nave. 'Thus', said the *ILN*, 'the two engineers who have spanned the Menai Straits . . . sleep side by side'.

45 The Camden Town railway, 1851

In November 1851 an *ILN* artist took a second-class return ticket (six pence) to Camden Town on the newly opened railway from Fenchurch Street. The first part of the trip was on the existing Blackwall Railway, through a covered way built 'to prevent horses taking fright at the noise and smoke of the engines'. Having passed the sugar-baking district of Goodman's Fields and the dock area he arrived at Stepney where passengers 'began to breathe more freely, for we had left behind the region of smoke and gigantic chimneys'. Next they passed some thirty acres of ground 'beautifully disposed and ornamented with cypress, cedar and other trees' – the City of London and Tower Hamlets cemetery – followed by 'the extensive buildings of the City of London Union Workhouse which . . . has a palatial appearance'.

Passing onward, 'through the verdant fields, we came to the retired village of Homerton'. At Hackney several long ditches 'nearly covered with what we took to be weeds' turned out be the watercress supply of London. Next came the model prison at Pentonville, built only a few years before, but already surrounded with houses. Then followed the crossing of the Great Northern Railway (see illustration) 'and it was a curious sight to see a monster northern train, sixty feet below us, entering the tunnel running under . . . Copenhagen Fields'. Then on by viaduct to journey's end at Camden Town, crossing roads by tubular bridges like that at Menai.

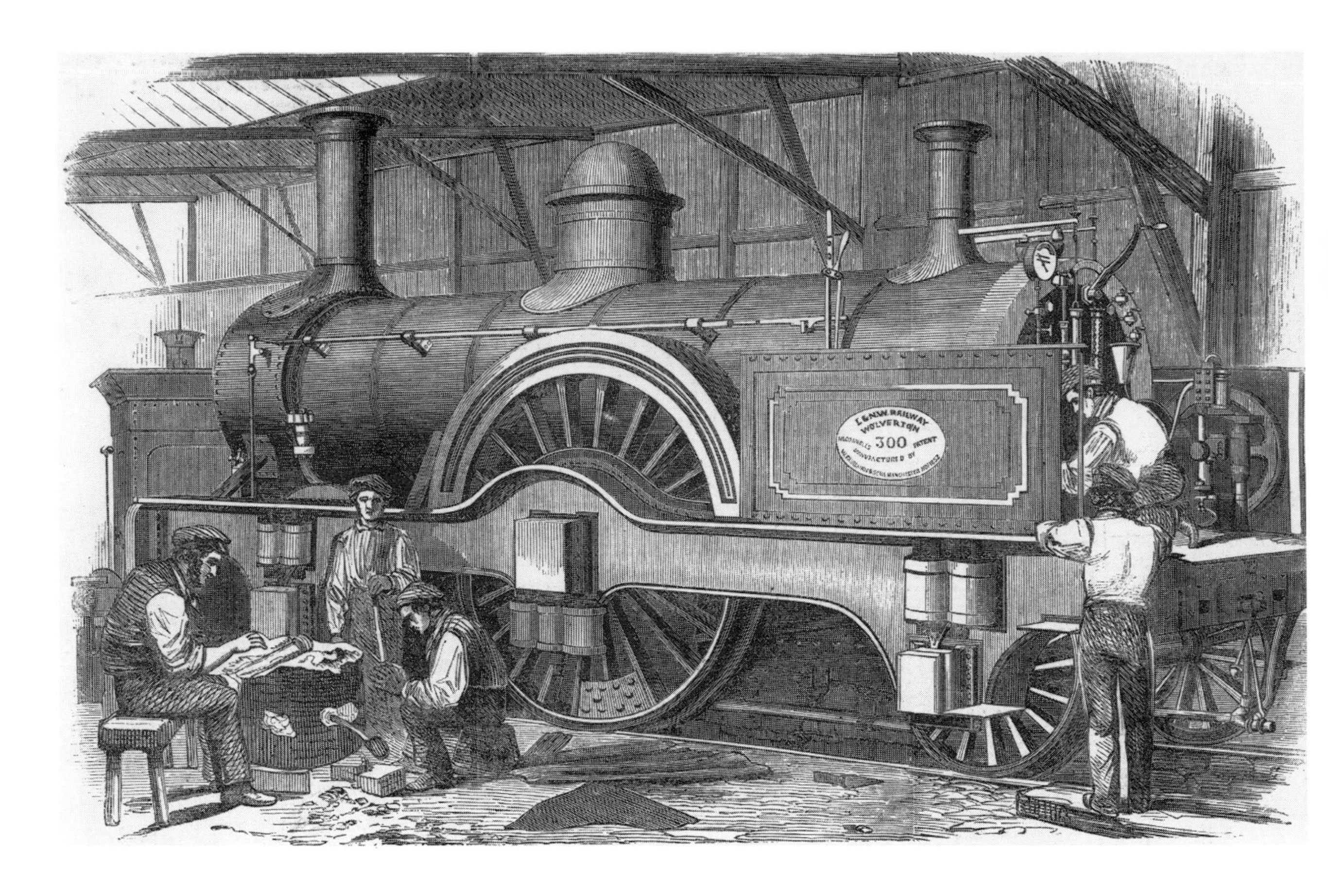

46 New express engine for the LNWR, 1852

The London and North Western Railway was formed in 1846 as a result of the union of the London and Birmingham, the Grand Junction and the Manchester and Birmingham Railways. At the time of the amalgamation there were locomotive workshops at Wolverton, Crewe and Longsight. The latter was closed in 1857. The one at Wolverton was closed for locomotive building in 1863; until 1861 its locomotive superintendent was J.E. McConnell, whom we meet elsewhere in this volume as the designer of the Marquess of Stafford's steam car (62). It was he who designed the '300' class of locomotive.

McConnell is reported as having pioneered experiments in which coal was used instead of coke as fuel. The '300' class incorporated a firebox and a combustion chamber which he designed and patented. The example illustrated was constructed by W. Fairbairn & Sons of Manchester.

47 The patent *Impulsoria*, 1850

The *ILN's* deadpan report of the arrival in London of a locomotive powered by horses is of the sort which modern readers only expect to encounter on 1 April. To make the vehicle move, four horses were yoked to its shafts. As they attempted to walk forward the moving belt on which they stood ran backward, and this motion drove the locomotive's driving wheels.

Described as 'lately invented in Italy', *Impulsoria* was in fact no more than an elaborated version of the *Cycloped*, a similar device whose inadequacy as a means of pulling trains had been publicly demonstrated at the celebrated Rainhill locomotive trials more than twenty years before.

Impulsoria was promoted on economic grounds, as a means of 'extending the advantage of the railway to locations hitherto impracticable'. There is no hint of a leg-pull in the reporter's suggestion that such a device might one day 'supersede the costly locomotive engine'.

48 Royal Albert Bridge, Saltash, 1859

In 1846 I.K. Brunel was faced with the problem of bridging the Tamar at Saltash. His final design, for a wrought-iron bridge with two spans of 455 feet and 17 short spans, met the Admiralty's requirements of not more than one pier in the fairway and minimum clearance of 100 feet at high water. It also met the railway company's demand for economy by providing only a single track.

The main challenge to Brunel was provided by the central pier. The water here was 70 feet deep and below it was 20 feet of mud and gravel. 'An immense wrought-iron cylinder, 37 feet in diameter, 100 feet high and weighing 300 tons was sunk exactly in the centre of the stream. From this, water was pumped out and air forced in until the men were enabled to work in comparative dryness at the bottom of the river in a kind of gigantic diving bell. By this means the whole of the sand and gravel was removed, the rock levelled, and the solid column of masonry reared from it to above the water line.'

The peculiar aesthetic virtue of this well-loved structure lies in its combination of the delicate mathematical elegance of the suspension bridge (now displayed in its purity by the adjacent road bridge of 1961) with the reassuring sturdiness of the tubular arches.

Writing in 1983, Bowden pointed out that whereas the bridge cost £225,000 in total to build, it now costs £320,000 just to paint.

49 The Workmen's Penny Train, 1865

As the railway companies drove their main lines into central London, there resulted 'the wholesale demolition of houses formerly inhabited by people of the working class'. In some cases the only compensation offered to these people was the sum of 1s 6d (7½p) towards the cost of moving. With a twinge of social conscience, Parliament insisted that the companies which caused the disruption should provide special services to convey 'artisans, mechanics, and daily labourers' between their places of work and the new suburban lodgings to which, it was hoped, they would have moved. The fare was to be a penny a journey.

The engraving shows the London, Chatham and Dover Company's portion of Victoria Station at a little before six on a March morning in 1865. Trooping up the platform are the honest working men – and at least two women – who were regular commuters from such south London suburbs as Camberwell, Brixton and Stockwell. The return service left at 6.15 p.m. or 2.30 p.m. on Saturdays.

The workmen's service was well patronized, though not in general by the class of person for whom it was instituted. Too poor to support a commuting way of life, these unfortunate refugees often resettled in even more cramped quarters close to the sites of their former homes.

50 The first locomotive into Indore, 1875

The first rail service in Asia was inaugurated at 3.30 p.m. on 16 April 1853, when a fourteen-carriage train steamed out of Bori Bunder on the west coast of India. The event was marked by a public holiday, a twenty-one-gun salute, and the applause of a vast multitude. The Governor, Commander-in-Chief, Presidency staff and Bishop did not attend, however; according to the London *Globe* they 'had left for the hills a short time before, not considering the most memorable event that ever occurred in India worthy of their presence'.

The line in question was a 24-mile fragment of the projected Great Indian Peninsular Railway, whose first directors had included Sir Jamjetsee Jeejeebhoy, Baronet, and Mr George Stephenson. The GIPR grew rapidly in the years that followed, and by 1875 was part of a 6,500-mile network. In that year work began to link the GIPR with the town of Indore in Central India, 80 miles away, and it is the arrival of the first locomotive for this line that is shown in the illustration. The Maharajah of Holkar, who put up the capital for the new 'Holkar Railway', was present in person to welcome the locomotive.

51 Cannon Street Station, 1866

ILN readers never seemed to tire of statistics and superlatives. In separate articles in 1866 they learned firstly that the new Cannon Street Station would in its 'vast extent and capacity . . . far exceed every other building of its kind', and secondly that the bricks used in its construction, laid end to end, would stretch 3,835 miles.

With the new station the South Eastern Railway gained a foothold in the City of London. A shuttle service was established between Cannon Street and the West End station at Charing Cross, and carried three million people a year until the coming of the underground railway took the trade away. The overground rail link avoided the congestion in the streets between the City and the West End which was as bad then as it is now, and the third-class fare of two pence was a penny less than that of the horse bus. Long-distance passengers heading for Charing Cross were less well served: they had to cross the Thames three times before reaching their destination, since most services called at Cannon Street *en route*.

Cannon Street Station survived largely intact until 1941 when the roof was damaged by enemy action. Today a new office block straddles the platforms; only the twin towers and part of the brick wall remain from the original building.

W
E

52 The Niagara Suspension Bridge, 1862

Seven years before Blondin attempted his crossing of Niagara on a tightrope, Charles Ellet jun., an engineer, traversed the gorge in an iron basket suspended from a wire rope. His 'basket ferry', inaugurated in March 1848, was the first of many structures to link the Canadian and American sides of the Niagara Gorge. Its construction began, it is said, with Ellet offering a prize of five dollars to the first boy who could fly a kite to the opposite bank.

The Great International Railway Suspension Bridge was the work of John A. Roebling, later celebrated as the designer of New York's Brooklyn Bridge. Roebling defied conventional wisdom which maintained that a suspension bridge could not be rigid enough for railway traffic. Stiffening was provided by trusses between the upper railway deck and the carriage and footway below, and by wire-rope stays linking the deck both to the support towers and to the rocks beneath.

The first locomotive steamed over with 'no vibrations whatever' in March 1855 and trains continued to cross the bridge – always at less than 5 m.p.h. – until 1898 when it was reinforced by converting it into a steel arch. This was destroyed by a build-up of ice in the gorge in January 1938.

The engraving follows a painting by the Danish-born artist Ferdinand Richardt in which the details of the bridge were themselves copied from a daguerreotype.

53 The projected Midland Grand Hotel, 1871

When the Midland Railway was completing its line into London in 1867 its directors decided to reward themselves by providing a terminus and hotel of unusual splendour. Three thousand slum dwellings were swept away to make room, and St Pancras station constructed. For nearly a century it would boast the largest station roof without central support in the world.

The adjoining hotel was to be known as the Midland Grand. Its architect was the eminent and autocratic Sir George Gilbert Scott, master of the gothic style. For its construction, high-quality materials – stone, iron and sixty million bricks – were brought from places along the Midland's routes. When the hotel opened in 1873 patrons found that the interior fittings were technically advanced as well as plush. There were gasoliers for lighting, and electric bells; the floors were fireproof, and residents could travel to the upper floors in a hydraulically-operated lift known as the 'ascending room'. In the kitchens, vegetables were prepared in iron steam-chests complete with safety-valves, while in the laundry a six-foot-diameter washing machine and two steam mangles were available to deal with three thousand pieces of linen each day.

Declining patronage led to the closure of the Midland Grand Hotel in 1935, though the building still stands.

54 St Gotthard tunnel – the first train, 1880

In the mid-nineteenth century tunnelling through the Alps must have seemed impossible to many, owing to the hardness of the rock, the impossibility of sinking shafts from the surface for ventilation and spoil removal, and the probable ambient temperature to be encountered. Nevertheless the problem was solved by the engineers who between 1857 and 1870, at a cost of twenty-eight lives and £3 million, joined the French and Italian railway systems by the 7.9 mile link that we call the Mont Cenis tunnel. The solution was made possible by the development of air compressors to operate the drilling machinery for placing the explosive charges, and to provide ventilation near the face.

After this triumph, the building of the St Gotthard tunnel (1872–82) to link Zürich with Milan was a long drawn out tragedy. Louis Favre, the contractor, had at his disposal improved water turbines and compressors, and the newly introduced explosive dynamite. But he had saddled himself with a suicidally low contract price which made him drive himself and his workmen literally to death. The cost of the 9.3 mile tunnel was 310 killed, 877 seriously injured, £2.3 million and the bankruptcy of the contracting company.

The illustration shows the first train (drawn by a compressed-air locomotive) to pass through the tunnel in 1880.

55 The Tay Bridge built, 1878 . . .

Thomas Bouch (1822–80) derived from an enlightened village schoolmaster an early interest in engineering. After nine years of experience he was, at the age of twenty-seven, appointed engineer of the Edinburgh and Northern Railway. Becoming acutely aware of the frustrations produced by the estuaries of the Forth and the Tay, he designed the 'floating railway' whereby goods wagons were ferried without unloading. After work on tramway lines, and on many short railway lines which included four important bridges and viaducts, it was as a very experienced engineer that he began work on the Tay Bridge in 1870.

The result was a structure two miles long with 85 spans. After inspection by the Board of Trade it opened on 31 May 1878. On 20 June 1879 the royal train carrying Queen Victoria south from Balmoral passed slowly over the bridge while the training ship HMS *Mars* fired a royal salute from the river below.

56 . . . and destroyed, 1879

At 7.13 p.m. on Sunday 28 December 1879, a train carrying seventy-five passengers, some on the last stage of their journey from Edinburgh to Dundee, left St Fort in the darkness to cross the Tay Bridge. A very strong gale was blowing down the estuary. Shortly afterwards telegraphic communication across the bridge failed; so also did the water supply of the village of Newport, which was carried across the bridge. The train did not arrive at Dundee, and when conditions made it possible to investigate, it was found that all three sections of the 'high girders' of the bridge had fallen into the river and with them the train.

The enquiry was conducted by W.H. Barlow, civil engineer, Colonel Yolland, Inspector General of Railways, and H.C. Rothery, Wreck Commissioner. Separate reports were issued, by the engineers and by the Wreck Commissioner. One essential difference between them was that the engineers were reluctant to blame specific individuals. The Wreck Commissioner had no such scruples. Bouch was pilloried with the words 'this bridge was badly designed, badly constructed and badly maintained'; the contractor was blamed for inadequate supervision of the foundry work; and the railway company was criticized for its practice of running trains across the bridge at speeds greater than those recommended in the inspectors' report.

57 The Metropolitan Underground Railway, 1863

The bypass road is not a twentieth-century phenomenon; London was bypassed in 1756 by the New Road, now split into Marylebone, Euston and Pentonville Roads. Along this highway were built the first great railway termini, Paddington, Euston and King's Cross. Separate schemes to link Paddington with King's Cross, and King's Cross with the City coalesced into a proposal to construct an underground Metropolitan Railway, and this was incorporated in 1854. Major financial support was supplied by the City Corporation, and the Great Western Railway's contribution of £175,000 ensured that the track, when laid, was of mixed gauge. The line was opened in 1863.

'The intermediate stations', said the *ILN*, 'form conspicuous and rather remarkable objects along the route'. The comments on these stations in Menear (1983) are tantalizingly brief. He points out that the two stations at each end of the line were in open cuttings with glass roofs over the platforms, but at Baker Street, Portland Road (now Great Portland Street) and Gower Street (Euston Square) there was a genuine underground feel about the stations with their brick-arched ceilings. Looking at the *ILN* illustration, mostly of exteriors, one sees that Baker Street and Gower Street were almost twins – and how one regrets the loss of the Portland Road domes!

58 The widened lines, 1868

On the Metropolitan from Paddington the section between King's Cross and Farringdon Street (starting with the Clerkenwell tunnel) was originally used also by Great Northern trains. When the line was extended to Moorgate it was decided to give the Great Northern an independent route to that station. This was done by duplicating the Clerkenwell tunnel by means of another to the north of it. Great Northern trains descending an incline through this new tunnel passed under the Metropolitan and then ascended to run side by side with it to Moorgate.

The illustration shows a Metropolitan train from Moorgate about to enter the original tunnel on its way to King's Cross, while at the lower level a Great Northern locomotive, having just left the new tunnel, is on its way to Moorgate. The group of four lines together became known as the 'widened lines'.

THE METROPOLITAN UNDERGROUND RAILWAY
PADDINGTON JUNCTION.
CHAPEL ST EDGEWARE RD
BAKER STREET.
PORTLAND ROAD.
GOWER ST EUSTON RD
SIGNAL MAN'S STATION AT KING'S CROSS
KING'S CROSS, INTERIOR
SIGNAL MAN'S STATION KING'S CROSS INTERIOR
KING'S CROSS, EXTERIOR
FARRINGDON STREET

59 The Forth Bridge completed, 1889

Before the collapse of the first Tay Bridge in 1879 its designer, Sir Thomas Bouch, had started to build a suspension bridge over the Forth. This was stopped and the present splendid steel cantilever structure was initiated in 1882. The engineers were Sir John Fowler and Mr (later Sir) Benjamin Baker; the contractors were Tancred, Arrol and Company. The work was completed in 1889 at a cost of £3.2 million and fifty-seven lives. The bridge carried two lines of railway track across two main openings, each made up of two 680-foot cantilever arms and a 350-foot suspended span.

It is often the case that the small details of great events are the most memorable, and readers who learnt at school about the thermal expansion of metal may be interested to read of a minor crisis that occurred when the two halves of one of the suspended spans, built out from the cantilever arms, were about to be bolted together. Owing to delays this was taking place later in the year than had been planned; the hoped-for ambient temperature was not reached and the relevant bolt holes remained 1/4 inch 'blind'. The day was saved by setting on fire 120 feet of wood shavings and oily waste soaked in naphtha, suitably placed to produce the required expansion of the booms.

60 Pullman dining car, 1879

The first Pullman sleeping car to appear in England was the *Midland* (1874), for use by the railway of that name. The Great Northern followed suit. Its first car, the *Ohio*, began service in 1875. Its third, the *India*, was eventually wrecked in the Thirsk railway disaster of 1892 (see 61). But the *Ohio* made gastronomic history in 1879, when it was converted into the dining car shown here, called the *Prince of Wales*, to a design by Waring.

The car was for first-class passengers, who paid a supplement of 2s 6d. Leaving Leeds for King's Cross at 10.00 a.m. on a four-hour journey and King's Cross for Leeds at 5.30 p.m. it allowed day-return businessmen time for a leisurely lunch and dinner. According to the *ILN* 'irregular dining hours have shortened the lives of many prosperous and active men of business who were little past middle age'; hence it welcomed the enterprise of the Pullman Car Co.

61 Railway accident at Thirsk, 1892

'There are few safer places on earth than a passenger compartment on an English train', declared L.T.C. Rolt at the beginning of his book *Red for Danger*, in which he mentioned 111 railway accidents that occurred between 1830 and 1940, and described in detail fifty-seven of them. Writing without sensationalism, he proved that pure accident was rare; almost always human fallibility was involved.

Every incident was investigated by ex-Royal Engineer officers of the Railways Inspection Department. The safety measures they recommended were usually accepted by the railway companies under pressure of public opinion, but sometimes legislation was needed. Thus together with the technology of railway operation there grew up a technology of railway safety. Its chief nineteenth-century achievements are summed up in the words 'lock, block and brake', i.e. interlocking of points and signals, preservation of space intervals between trains, and use of brakes (applied to every wheel) which acted automatically if the train became divided.

The illustration is of an accident near Thirsk in which the second part of an Edinburgh–London night express ran into the back of a goods train. It was caused by a signalman who fell asleep. Charged with manslaughter, he was found guilty but immediately discharged. The verdict was greeted with cheers because throughout the day and night preceding the accident he had been dealing with the illness and death of his child; he had sought relief from night duty but failed to obtain it.

62 Steam carriage for common roads, 1860

In 1950 R.W. Kidner wrote a book called *The First Hundred Road Motors*. His first motor was dated 1788 and his one hundredth 1871, and all were powered by steam. Number sixty-six, dated 1859, was designed by J.E. McConnell, Locomotive Superintendent, Southern Division, London and North Western Railway. It was built by Thomas Rickett of the Castle Foundry, Buckingham, and driven (with the assistance of a stoker) by no less a person than the Marquess of Stafford who, two years later, became the third Duke of Sutherland. In the illustration the nobleman is seen driving his 2½ ton three-wheeler to Windsor Castle for inspection by Queen Victoria and the Prince Consort. Stops every ten miles or so were needed, since in that distance the engine used up the whole of its 90 gallons supply of water.

A year later Rickett supplied a second and improved version of his vehicle to another peer of the realm, namely the Earl of Caithness. Dramatic accounts survive of an epic three-day ride by the Earl, his Countess and his minister, from Inverness to their castle near John O'Groats, a distance of some 150 miles. The climax came at the Ord of Caithness, 'a noted mountain' some 20 miles south of Wick with occasional gradients of 1 in 7. 'Ye'll ne'er get over the Ord!' said the locals, but they did get over it and the cannon of Wick sounded for them on the other side.

63 Clifton Suspension Bridge opened, 1864

The Clifton Bridge originated in the will of William Vick (died 1754), wine merchant, whereby £1,000 was to be invested until it had grown to £10,000 when it would be used to bridge the Avon gorge. In 1829, with the fund at £8,000, a design competition was held, which called forth twenty-two entries, with one by I.K. Brunel, then aged twenty-three. All were rejected by the judge, Thomas Telford, whose own design, involving a cautiously short central span and piers occupying the full depth of the gorge also proved unacceptable. A second competition was held, and this was eventually won by Brunel with a design involving a central span of 702 feet and chains passing over two towers supported on the rocky edges of the gorge.

Building began with due ceremony in June 1831, but adverse circumstances, including financial stringency, checked it for five years. A fresh start led no further than the completion of the two towers, and when work stopped again the ironwork was sold, the chains going to the South Devon Railway for incorporation in Brunel's Royal Albert Bridge at Saltash (48). Brunel died in 1859, but a company formed in 1861, with William Barlow and John Hawkshaw as engineers, completed this most elegant of bridges as a memorial to him, using chains from his own Hungerford Bridge, then being dismantled to make way for Charing Cross railway bridge.

64 Holborn Valley viaduct as projected in 1867

Horses and donkeys will caper like fleas
No more sore shoulders and broken knees
The Animal Society may take their ease
Goodbye to the once Holborn Hill.

In the early nineteenth century, traffic leaving the City of London for the west, having descended into Farringdon Street (which occupied the valley created by the Fleet River) had to climb out of it. Then could occur 'those terrible scenes of distress to the horses that . . . toiled up those steep ascents, goaded and flogged, to the disgust of every person who possessed the slightest feeling of kindness and tenderness'.

Hence the above extract from a sheet of doggerel that greeted the opening by the Queen on 6 November 1869 of the Holborn Valley viaduct. This was an almost level road that ran for a quarter of a mile through the City, crossing Farringdon Street by a three-arch skew bridge of cast iron. It was engineered by William Haywood and cost £2.1 million. The substructure was 'a perfect honeycomb work of arches; there are arches for the subways, arches for the sewer, arches for the cellars of the houses on either side, arches below the cellars and arches again below the arches, the approaches to which are on the low level of Farringdon Street'.

It was not all gain; some valuable property was 'improved out of existence', and Saint Andrew's church lost much of its churchyard, about two thousand of the residents being transferred to Ilford Cemetery.

65 Electric tramcar, 1883

The first public electric tramway was that built by Siemens at Lichterfelde, Germany in 1881. Where the line ran along the street an overhead conductor was used; elsewhere the current passed down one rail and returned by the other, and here the track had to be fenced. But just at that time the electrical 'accumulator' had been invented and developed in France and there arose the possibility of a tramcar carrying its own medium-voltage power supply, and riding on an unfenced track. Hence great interest was shown when Anthony Reckenzaun of the Electrical Storage Company developed such a tramcar and arranged to show it, first to General Hutchinson, Board of Trade Inspector, for approval, and then to a party of some forty VIPs, including the Governor of the Bank of England, assembled at Kew Bridge on 10 March 1883.

All went well in the Board of Trade test from Shepherd's Bush to Acton. But on the return journey the driving belt kept on slipping and eventually the car had to be drawn by horse to Kew and repaired while the impatient VIPs were soothed by giving them trips in Reckenzaun's electric boat. At last the party set off for Gunnersbury, escorted by the horses, which had to be attached at every upward gradient, and also when the motor failed to start after a stop. The periodical *Engineering* was severe on this fiasco. 'It is a pity', it said, 'that electric traction should have been made to look so ridiculous'.

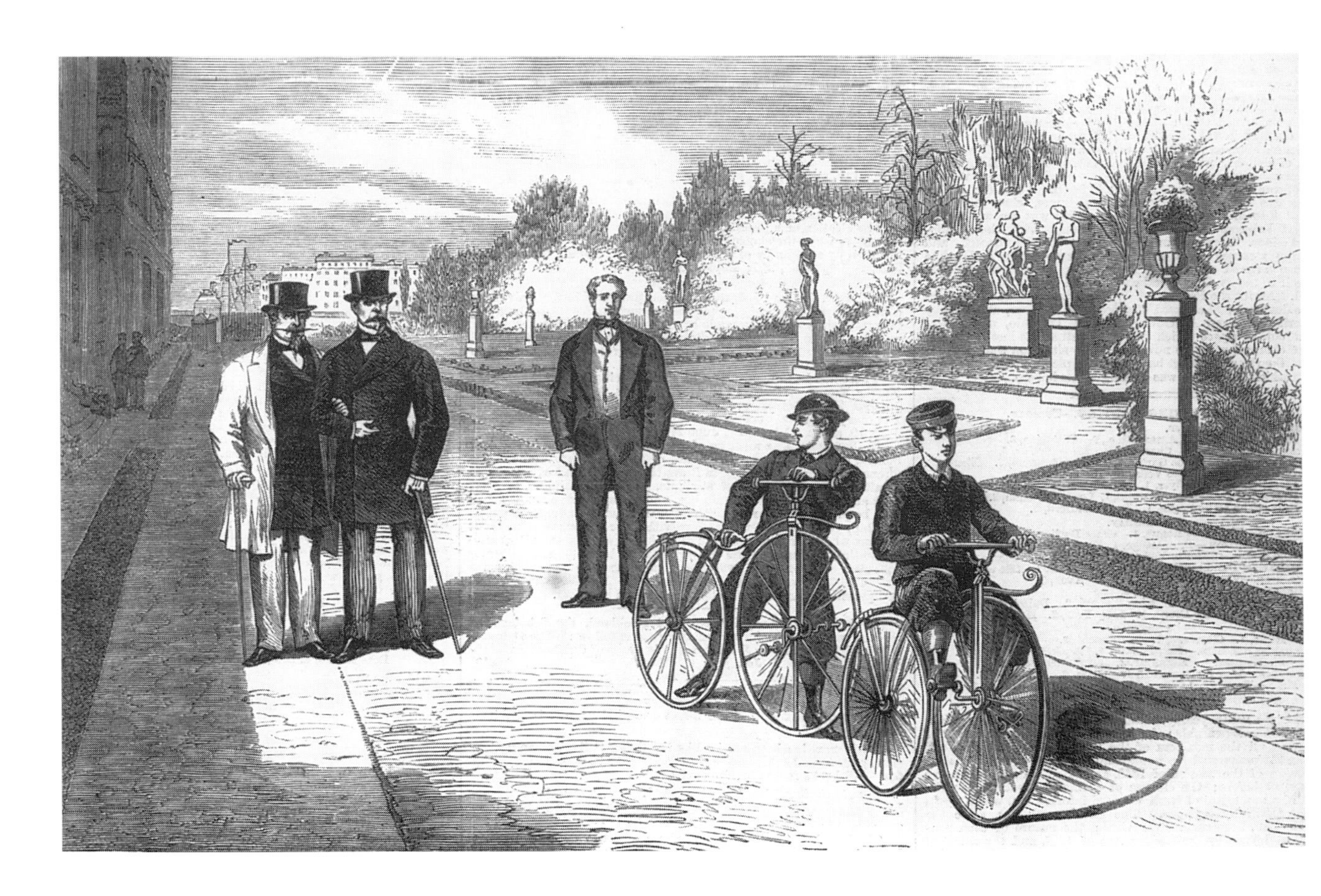

66 Imperial velocipede, 1869

The bicycle, surely man's most benign and blameless mechanical invention, began life in 1817 at Mannheim as Drais' *Laufmaschine* (running machine). Its rebirth (1861) as a machine pedalled at the front wheel was due to Michaux of Paris, whose device was named the *vélocipède* in France and nicknamed the 'boneshaker' in the United Kingdom.

In the illustration, the place is a private part of the Tuileries gardens; the personnel are the Emperor of the French, his aide General Fleury, his son the Prince Imperial, Conneau the Prince's playmate, and an attendant. The Prince has already learnt to manage his mount 'with tolerable ease and precision'. He takes the lead, of course, and 'delights his father with his display of precocious skill'. Conneau, however, is rather behindhand 'unless it be that he has already learnt to be a courtier'. The attendant is there to cope with a possible tumble.

67 The cyclist's camp at Alexandra Park, 1884

An over-hasty glance at this illustration and its text suggests that it represents a camp at Alexandra Park of military volunteers, i.e. of men prepared in an emergency to put their cycling skill at the service of the army. The more ambitious of such volunteers hoped that they might be employed as cavalry, while the more realistic expected to be occupied at best as scouts and at worst as messengers. In fact the illustration represents a privately promoted Whit Monday meeting for members of cycling clubs in the London area. The day was spent in racing while in the evening there was a torch-light procession followed by camp fire activities.

The *ILN* wrote enthusiastically about this event, describing it as 'one of the greatest sights for Londoners on Bank Holiday'. But *The Cyclist* regarded it as a flop, suggesting that right-minded London cyclists wanted to get out of town to (say) Hindhead on a bank holiday, while country cyclists would like to 'plunge into the vortex of London excitements and not practice sham rusticity at Hornsey'. Sympathy was expressed for the financial loss incurred.

The illustration is interesting in showing that in 1884 the tricycle was not, as it is today, a vehicle for the eccentric or the aged. Tricycles form the majority here. The 'safety bicycle', though it had appeared by 1884, is not represented.

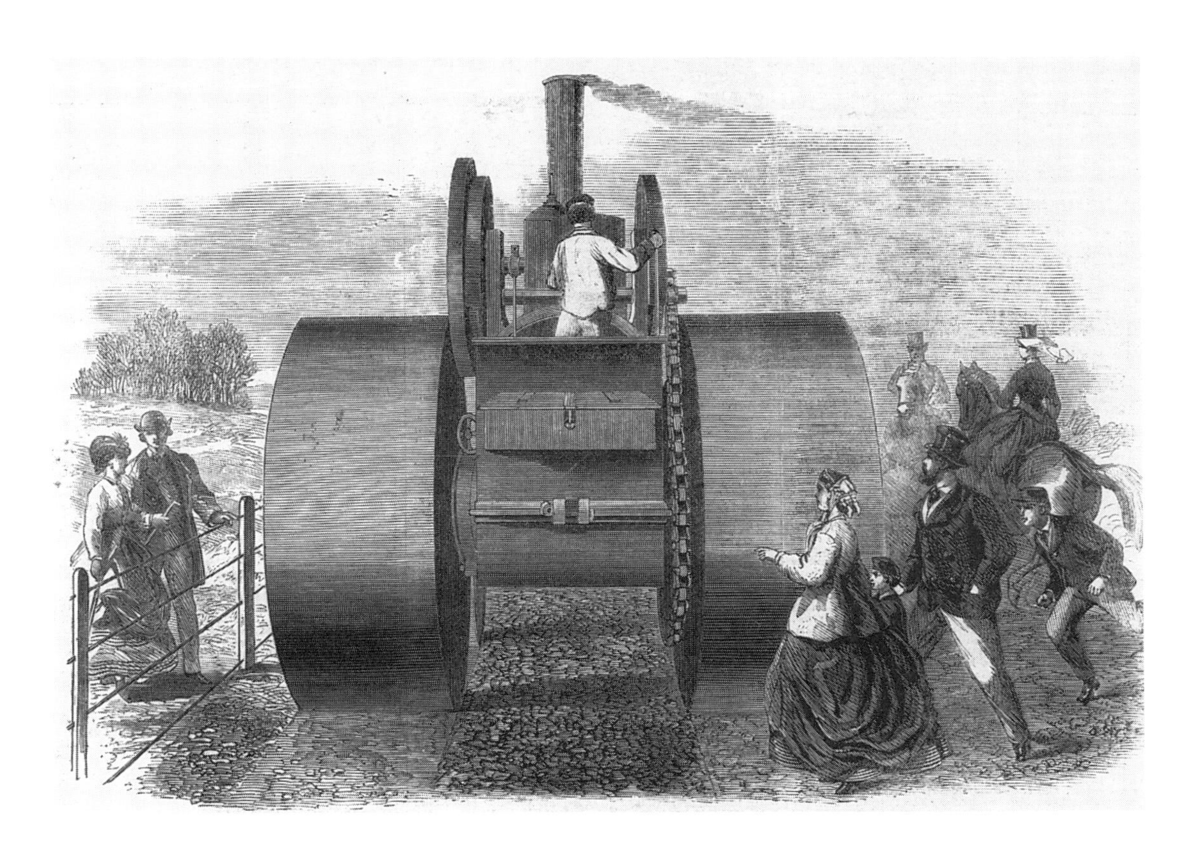

68 Steamroller, 1866

By the mid-nineteenth century the steamroller was calling out to be invented, since it might be expected to perform with ease a task which could not readily be done by horses. To compact a macadamized road the limit of effort available was a 5-ton roller drawn by six horses; a bigger team, even if efficiently controllable, would do more damage to the surface than could be repaired by the roller. A similar problem arose if the roller was dragged by a tractor.

A solution appeared in the form of an integrated steamroller and on 15 December 1866 the *ILN* produced a rear view of this novel machine, manufactured by Aveling and Porter, at work in Hyde Park. The engine was of 12 horsepower and the rollers were three feet broad and seven feet in diameter 'bearing with a weight of three tons on each square foot of road'.

69 'Who wants a motor-car?', 1899

Not, it would appear, the trilby-hatted lady in the forefront of this picture, since the accompanying verses read:

Who wants a motor-car?
Though wives may storm and daughters weep,
Yet purses short and prices steep
Make motors much too frail to keep
So long as trains are swift and cheap:
Who wants a motor car?

No maid who has her wheel:
She loves to curl, and twist and wind,
She loves the cart-ruts intertwined,
And, if the Fates are very kind,
She loves the man that rides behind
And drives the skidding wheel.

For those who miss the message conveyed by this deplorable verse the essence of the matter is 'No girl who rides tandem with her boy friend behind, wants a car'.

The interest of the illustration, which appeared in the last month of the nineteenth century, lies in the variety of vehicles shown, including a delivery van. One notes with regret cyclists in the fast lane obstructing the motorists.

The Queen opens the Great Exhibition of 1851

Science: Created, Explained, Applied and Exhibited

Natural philosophy is not easily illustrated; its creators are. Our portrait of Faraday appeared in the *ILN* in 1861, the year of his retirement.

The *ILN* worked nobly to further public understanding of science, in particular by reports of lectures at the Royal Institution. We include a picture of Tyndall speaking there. Activities at the Royal Polytechnic were also featured and Pepper's ghost was raised twice in the *ILN*.

Under astronomy we honour that unsung hero of the art, the lonely observer. Perhaps the most spectacular of the illustrations of *applied* science is that associated with radiology. An early adventure of Big Ben symbolizes horology. Associated with early photography we read of a man who was nearly arrested for taking a shot of, not at, the Queen. Acoustics is worthily represented by a splendid portrait of the great Thomas Edison with his favourite invention.

Science, both pure and applied, formed one side of the activities of the South Kensington Museum; we recall an important loan exhibition that was held there in 1876. Scientific instruments featured in the Great Exhibition (opened 1 May 1851; see opposite) and in the International Exhibition of 1862. In the former the most commanding exhibit was the Crystal Palace itself.

70 Michael Faraday, 1861

Faraday was born in 1791, the son of a blacksmith. After apprenticeship to a bookbinder, during the course of which he gained some knowledge of science by reading and attending lectures, he contrived to become assistant to Sir Humphry Davy, Professor of Chemistry at the Royal Institution, and between October 1813 and April 1815 toured Europe with him. Returning to the Royal Institution he became Director of the Laboratory from 1825 until retirement in 1861.

From achievements unparalleled in number and potentiality we select six. Faraday demonstrated the identity of electricities of different origin; he established the laws of electrolysis; he produced continuous rotation by means of an electric current; he discovered electromagnetic induction; he discovered diamagnetism; and he introduced the highly fruitful concept of the 'field' for the interpretation of electric and magnetic phenomena.

By his lectures, especially those for juvenile audiences, he became known to many. To quote the *ILN*, 'the "juveniles" are worth seeing; they include princes of the royal blood, savants of European reputation, future dukes and earls, fair ladies and grey-haired professors, intense students animated with a love of science, chemists, doctors, poets, lawyers and wits; these usurp the places of boys, mix with them and crowd them on their seats. We are all juveniles when we hear Faraday'.

The man whom Tyndall, referring to his science, described as 'this mighty investigator' and, referring to his character, as 'this just and faithful knight of God', died on 25 August 1867.

71 Mr C.R. Darwin FRS, 1871

On 15 January 1871 Charles Darwin (1809–82) returned the final proofs of his book *The Descent of Man* to his publisher, John Murray. Two days later he started work on another book.

The Descent of Man was published in an edition of 2,500 copies at the end of February. To mark the occasion the *ILN* provided a full-page portrait of the author, from a photograph by Ernest Edwards, supported by a 3,500-word essay on his life and work. Darwin had expected that the new book would, like its predecessor *On the Origin of Species*, be given a stormy reception. The *ILN* article, if he read it, would have relieved his anxiety, for it gives a balanced, if sceptical, critique of the theory of evolution and the way it had been received in Victorian Britain.

Mr Darwin is a 'naturalist and philosopher', we are told, whose *Origin of Species* was a 'bold and ingenious essay'. It 'has been vehemently abused . . . by illogical and intemperate partisans . . .' reacting with revulsion to the suggestion that man is closely related to 'that detestable creature, the ape.' The hypothesis of the new book, that the ancestry of man can be traced back to 'a group of marine animals, resembling the larvae of existing Ascidians', seems less repugnant. Mr Darwin's theory, concludes the *ILN*, is not inconsistent with the concept of a 'Divine power and wisdom in creation', but in the end 'we must leave the subject to thoughtful readers'.

72 Professor T.H. Huxley FRS, 1870

Professor of Natural History in London for thirty-one years and honorary doctor of eight universities, Thomas Henry Huxley (1825–95) is best remembered not as a researcher but as a popular exponent of the principles of natural philosophy. He was well known to lay audiences at the Royal Institution, the South Kensington Museum, and elsewhere. 'The exquisite clarity of his style, whether in speaking or in writing . . . won him the delighted attention of thousands who knew next to nothing of the "ologies" and the "ics".' As explainer and promoter of the theory of evolution Huxley was described as the man who hatched the egg that Darwin had laid.

The portrait, from a photograph by Elliot and Fry, marked Huxley's presidency in 1870 of the British Association for the Advancement of Science. In his presidential address he discussed the genesis of life, saying it was a subject whose history repeatedly revealed 'that great tragedy of Science – the slaying of a beautiful hypothesis by an ugly fact'.

73 Sir Richard Owen FRS, 1892

Richard Owen (1804–92) was born in Lancaster and educated for a medical career, but chance led him into research in comparative anatomy, vertebrate palaeontology and geology. This culminated in his appointment in 1856 as Superintendent of the Natural History Departments of the British Museum. He published some 650 articles and books. A list of his distinctions contains ninety-two items. From the Queen he received in turn a civil list pension, the use for life of a cottage in Richmond Park, and a KCB.

Despite this success, Owen was in old age sadly aware that advancing science had made much of his work irrelevant. However, he could have had no conception of the enormous importance to popular science today of his 1842 contribution to the Report of the British Association. In this he classified *Iguanodon, Megalosaurus* and *Hylaeosaurus* as *Dinosauria* and so unwittingly founded the 'dinosaur industry'.

74 Pepper's ghost, 1863

Before 1863 the lot of the stage ghost was not a happy one. It could appear, deliver its lines (if any) and vanish. Anything more ambitious would reveal its corporeality. After 1863 it might walk across the stage, be stabbed or shot at and depart unscathed through an unopened door. All this was made possible by apparatus consisting essentially of a vertical sheet of plate glass set diagonally across the stage, some screens, and the necessary illumination. The basic principle was developed by one Henry Dircks, who demonstrated it at a British Association meeting in Leeds in 1858. He could not persuade a theatre to take it up, but a model was marketed as a toy. One of these came into the hands of 'Professor' Pepper (1821–1900), the great Victorian showman of science, and he exploited the technique, initially at the Royal Polytechnic, of which he was then Director.

A joint patent was secured and Dircks (who had independent means) made over to Pepper all profits from self-promoted or licensed performances, a concession worth thousands of pounds. But the partners soon fell out. In his book *The Ghost* (1863) Dircks ascribed this to the failure of Pepper to give credit to him as inventor in the publicity. Pepper's account did not appear until 1890; in it he played down the controversy.

75 The Education Collection at the South Kensington Museum, 1857

The South Kensington Museum (parent of the Victoria & Albert Museum and the Science Museum) opened in 1857 in a modest building which was called by authority the Iron Museum and by the public the Brompton Boilers. The building was provided by the Commissioners of the 1851 Exhibition, and the Department of Science and Art, which administered the Museum, showed not a little ingratitude when in its fifth report (1858) it informed Her Majesty that the building was 'unsuitable in every aspect for the conservation of objects of value'. It was too hot in summer and too cold in winter, the roof leaked, and the fire risk was considerable.

The museum thus ignobly housed contained nine collections, one of which comprised items such as models, diagrams, and books, used in education. It presented 'a forceful illustration of the readiness of the public to co-operate with the state in promoting public objects, the state simply finding house room and superintendence, whilst publishers and producers of educational books and apparatus, both at home and abroad, are willing to offer contributions'. The collection was, in fact, a trade show and a thoroughly good and useful one, which enabled 'all persons engaged in teaching to see, collected together in one group, the most recent, the best and the cheapest forms of apparatus and means of imparting knowledge'.

The illustration gives a good general impression but is weak in detail and, apart from the globes, the only recognizable piece of apparatus is the Astronomer Royal's model of the Greenwich transit circle (see 82).

76 Tyndall lecturing, 1870

John Tyndall (1820–93), natural philosopher, teacher, lecturer, scientific entertainer, public examiner, consultant, spokesman for science, mountaineer, glaciologist and controversialist was above all the king of demonstrators. His throne was at the Royal Institution, with which he was associated for thirty-four years, at first as Professor of Natural Philosophy and later (in succession to Faraday) as Superintendent of the House. He had ample opportunity to display his regality since his appointment required him to give nineteen lectures during the season and two Friday evening discourses. The high density of demonstrations associated with such performances is illustrated by the fact that in one of them he carried out about thirty meticulously rehearsed experiments on vibrating strings.

In a memoir on the deceased Faraday, Tyndall used the words 'nothing could equal his power and sweetness as a lecturer'. Precisely the same phrase can be used of Tyndall himself, and we need not feel coy about employing the word 'sweetness' since Thompson, comparing Huxley (72) and Tyndall, says that 'where Huxley forced the scientific pill down his victims' throats, Tyndall wrapped it up neatly, coated it with sugar and induced them to accept it voluntarily'.

77 The air walker at Drury Lane theatre, 1853

The air walker was Mr Sands, equestrian and proprietor of the Hippodrome at New York. To perform his act he laced sandals over his boots, and to them attached brass loops, connected by springs to a pair of platter-like soles. He then climbed a ladder to one of the slung seats, and 'lying on his back by the aid of the ropes, placed his platter-shod feet on the ceiling, gently detached himself, and very slowly walked across the platform, occasionally poising himself on one leg'.

Every schoolchild can calculate, assuming the equipment to be perfect, the enormous force that Sands would have to exert to pull one foot away from the ceiling, given that the platter had a diameter of 0.3 metres and the atmospheric pressure was 101,325 pascals. So how did he do it? The *ILN* suggested a valve, but did not venture into detail.

78 Big Ben – disaster and triumph

On 16 October 1834 the Palace of Westminster was largely destroyed by fire. One element of Mr (later Sir) Charles Barry's design for a new palace was a tower to hold a clock, a big hour bell and six smaller bells. The designer of both clock and bells was Edmund Beckett Denison, later Sir Edmund Beckett, later Lord Grimthorpe, who is described in the *Dictionary of National Biography* as lawyer, mechanician and controversialist. His talents in the last-mentioned capacity were fully stretched by his association with Barry; indeed, as the designer of his own house at St Albans, he described himself as 'the only architect with whom I have never quarrelled'.

Our first picture shows the big bell, cast by Warner & Sons of Cripplegate at their foundry at Stockton on Tees, after its journey by rail, sea and road to Westminster. Here it was given a temporary mounting at the foot of the incomplete tower, and was tested by striking it with a six-hundredweight hammer for a quarter of an hour each week for ten months. Suddenly, on 17 October 1857, the big bell became flat. Inspection revealed a crack. Tempers became sharp, with designer and contractor at odds. The bell had to be recast and the contract was awarded to George Mears of the Whitechapel Bell Foundry, which claimed foundation in 1420.

Firstly the bell had to be broken up. It was laid on its side, and over two days an iron ball weighing about 2,700 pounds was repeatedly dropped on it. The wags asked if the new bell would be called Big John, since Lord John Manners had by then succeeded Sir Benjamin Hall as Chief Commissioner of Works. The recast was apparently successful but there was much more drama to come, accompanied by the rumblings of Denison-Barry discord, before the great clock began to serve London, a source of intense national pride.

DINING ROOM

79 Speaking tube, 1887

A short tube that is broad at one end and narrow at the other becomes a speaking trumpet (or megaphone) if the narrow end is held to the mouth, and an ear trumpet if it is held to the ear. If, however, a tube is long and of uniform diameter it is found to have a remarkable power of transmitting human speech with much less attenuation than might be expected.

In the nineteenth century the 'speaking tube' was much used for communication between upstairs and downstairs in great houses and between bridge and engine room in steamships. Later it was employed also between cabby and fare in taxis.

The *ILN* picture shows a watchman on the roof of the headquarters of the London Fire Brigade at Southwark. If he spots a fire he has a compass rose whereby he can tell its approximate direction and a speaking tube whereby he can raise the alarm.

80 Measuring the wind, 1880

The scene is the north-west turret above the Great Room of the Royal Observatory, Greenwich, and the time is late January 1880. Dramatis personae are two members of the staff of the Meteorological and Magnetic Division of the Observatory, who have come to service the self-recording instruments in the turret. It is cold, and strong gusty winds and sheets of rain are giving the instruments plenty to do.

Above the turret the vane is that of an Osler Anemometer by Newman. It is linked in such a way that the angular movement of the vane gives a horizontal movement to a clip, inside the turret, which holds a vertical pencil whose point touches a strip of paper. This strip is driven past the pencil by a clockwork motor and thus provides a continuous record of changes of direction of the wind. The circular plate above the turret is that of another component of the anemometer which records on the same strip of paper the pressure of the wind, and the structure on the left of the turret holds a pluviometer or rain-gauge which, in effect, weighs the rain continuously as it accumulates in a copper vessel.

The Astronomer Royal, G.B. Airy (81), had complete contempt for meteorology as a science, complaining that it lacked 'suggestive theory', but he was prepared to believe that regular observation might lead to useful empirical results.

81 George Biddell Airy FRS, Astronomer Royal, 1868

When Airy (1801–92) retired in 1881 the Board of Visitors of the Royal Observatory, Greenwich, selected as some of his greatest achievements the complete reorganization of the equipment of the observatory, the design of instruments of exceptional stability and delicacy, extension of the means of making observations of the moon, investigation of the effect of the iron of ships upon compasses, and the establishment of a national system of time signals.

McCrea described Airy as the greatest of the Astronomers Royal. Having listed his virtues he looked for his flaws and said 'perhaps the worst was that he had none'. The human warmth that so often accompanies fallibility might have reconciled his assistants to his rigid discipline.

The engraving is from a photograph by J. Watkins.

82 Night work at Greenwich, 1880

➸

To the Astronomer Royal, Airy, observing was the most menial of the tasks performed by the staff at Greenwich Observatory. 'No intellect and very little skill are required for it,' he wrote in 1847. 'An idiot, with a few days practice, may observe very well.' The observer in this engraving, whose intellectual capacity is not recorded, is using Airy's Transit Circle, a telescope constrained to turn only about a fixed axis which runs precisely east–west.

In his field of view the observer saw ten vertical lines, which were spider threads placed at the focus of the eye-piece. As the Earth turned, stars were seen to glide across the field. To make an observation the observer had to press an electric button at the precise instant at which a selected star passed each of the threads. This produced a trace on a moving chart which enabled the moment of passage of the star to be timed against the ticking of the Observatory's standard clock. If there was a discrepancy it was the clock which had to be corrected, for it is rotation of the Earth which ultimately defines the length of a day.

In 1880, Greenwich Mean Time, determined in this way, became the official time throughout Britain. Four years later the Greenwich meridian, which by definition runs directly under this instrument, was adopted as the world's prime meridian of longitude.

83 Refracting telescope for the Vienna Observatory, 1881

When in 1876 the government of Austria-Hungary decided to 'crown the observatory at Vienna with a refracting telescope of surpassing power', and to acquire the finest instrument that could be procured, they sent a representative to visit all the principal observatories and workshops in the world. He selected Mr Howard Grubb of Dublin for the task of constructing the instrument. The result was the 27-inch refractor shown in the illustration. When installed it was the biggest of its kind in the world but it enjoyed this distinction for only five years.

Howard Grubb was born in 1844. His father, Thomas Grubb, was engineer to the Bank of Ireland, but took up practical optics as a hobby, and eventually established a workshop from which emerged, as his greatest achievement, a 4-foot reflector for Melbourne. Howard Grubb took over and relocated the workshop in 1868, and from it came a stream of large telescopes including those for Cape Town, Oxford, Greenwich, Johannesburg, and Simeis in the Crimea. Grubb also carried out the very exacting task of constructing seven 13-inch photographic refractors for an international photographic survey of the heavens.

Grubb was elected FRS in 1883 and knighted in 1887. Wartime contracts on the submarine periscope that he perfected brought his business to St Albans, and finally after a take-over by Sir Charles Parsons it migrated to Newcastle in 1925. Sir Howard then retired and he died in 1932.

84 Photography with magnesium light, 1865

We are told that at the opening soirée of the meeting of the British Association at Birmingham in 1865 'photography by the magnesium light and the consequent production of that class of miniatures called "pistolgrams" was a source of some amusement and attracted large groups around the place of operation'.

The potentiality of burning magnesium as a source of light in photochemical experiments was revealed in 1859. Magnesium was then a very rare metal and it was not until 1865, when it could be produced at a price per ounce about double that of silver, that this illuminant became generally available to photographers. The first portrait using this technique had been taken the year before. It was of Henry Roscoe, chemist, by Alfred Brothers, photographer, both of Manchester.

The 'pistolgram' mentioned in the quoted report refers to one Thomas Skaife who in 1856 produced a pioneer miniature camera which he called a 'pistolgraph'. To quote Gernsheim (1969), 'when he aimed his "pistolgraph" at Queen Victoria he was nearly arrested for attempting to shoot her; unfortunately this photograph was lost forever when Skaife had to open his "pistol" to convince the police that his "shots" were harmless photographs'.

85 Eadweard Muybridge at the Royal Society, 1889

Eadweard Muybridge was born Edward Muggeridge in Kingston-upon-Thames in 1830, but changed his name when he emigrated to America as a young man. He began to make a reputation as an innovative photographer, but his career went awry when he murdered his wife's lover. Though acquitted by a California jury, he felt it best to disappear for a while, and spent some years travelling in Central America.

He returned in 1877 to continue a photographic investigation into animal locomotion which had begun when he was commissioned to find out whether a trotting horse ever has all four legs off the ground at once. The question cannot be resolved by the unaided eye, and was hotly debated in racing circles. Muybridge set up a battery of cameras to take a sequence of short-exposure photographs of a trotting horse, and so confirmed that 'unsupported transit' does indeed occur.

He later developed the zoöpraxiscope, a variant of the magic lantern in which the slide consists of a rotating disc with multiple images around its perimeter. Combined with stroboscopic illumination the device projects a rapid sequence of pictures which produce the illusion of movement on the screen. Demonstrated by its eccentric creator at the Royal Society's conversazione in London in 1889, this forerunner of the cinema projector excited much interest.

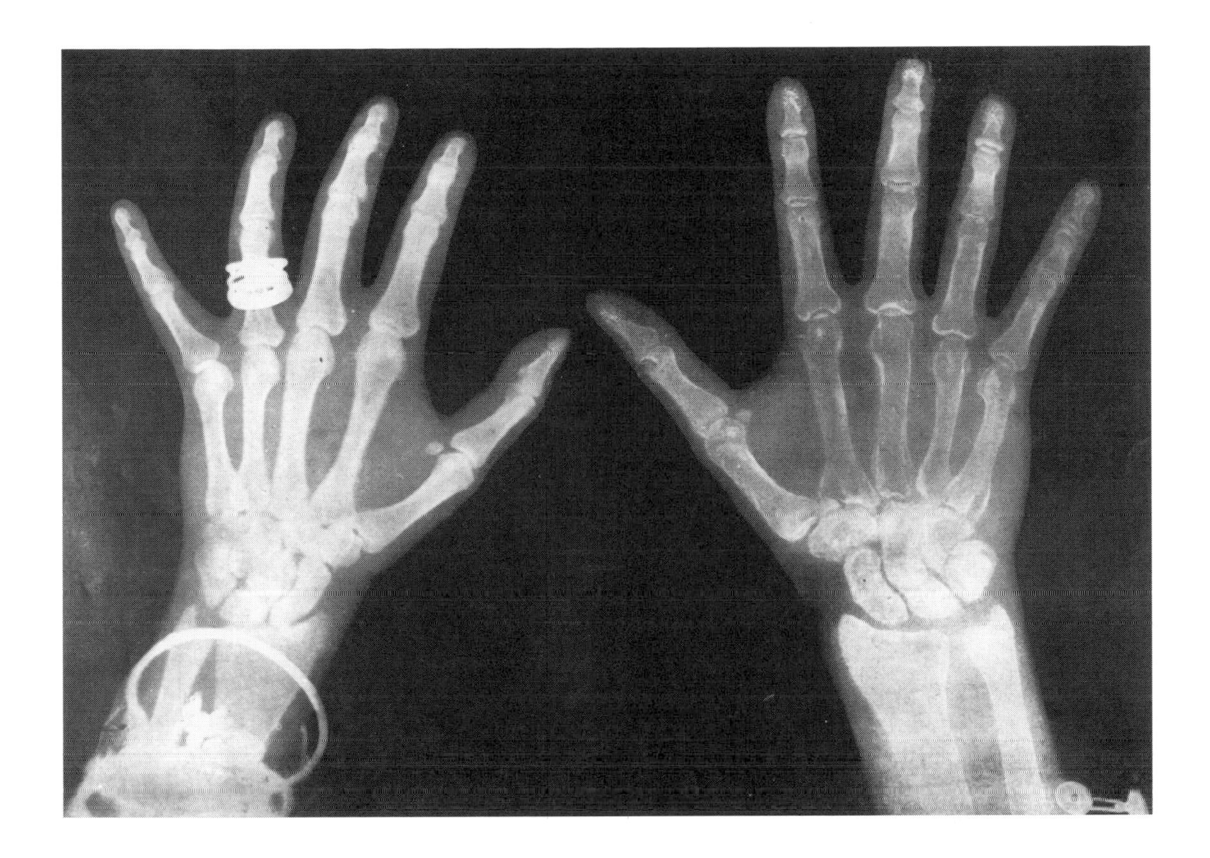

86 X-rays, 1896

In April 1896 an article was published by a reporter who had secured an exclusive interview with Professor Wilhelm Conrad Röntgen of Würzburg with reference to an important recent discovery.

'I was working with a Crookes tube covered by a shield of black cardboard,' said Röntgen. 'A piece of barium platinocyanide paper lay on the bench there. I had been passing a current through the tube and I noticed a peculiar black line across the paper.'

'What of that?'

'The effect was one that could only be produced, in ordinary parlance, by the passage of light. No light could come from the tube, because the shield which covered it was impervious to any light known, even that of the electric arc.'

'And what did you think?'

'I did not think; I investigated.'

Investigation must be guided by thought, so these words of the Professor, as he described his discovery of X- (or Röntgen) rays, have stimulated much discussion.

One of the hands in the illustration belongs to the Duke of York, the other to his Duchess (later King George V and Queen Mary). The *ILN* commented that 'the Röntgen ray is no respecter of persons, and gives a touch of homeliness to the most illustrious anatomy'.

87 Edison and the 'perfected' phonograph, 1888

Thomas A. Edison (1847–1931), the great American inventor, was the first to record and reproduce sound. In his tinfoil phonograph of 1877 the recording device was a stylus, mounted at the centre of a diaphragm, which vibrated under the influence of the sound waves generated by the speaker's voice. The medium was a sheet of tinfoil wrapped round a helically grooved brass cylinder. The sound track was a series of indentations made by the stylus along a helical path as the cylinder, rotated by hand, advanced beneath it. The reproducer was a stylus which was made to follow this track, and to communicate the resultant vibrations to a diaphragm.

Having created this device, with its enormous potential, Edison turned his attention elsewhere and it was left to Chichester Bell and Charles Sumner Tainter to produce, in 1885, the 'graphophone' in which evanescent indentations on a sheet of tinfoil were replaced by permanent incisions on a wax cylinder. The resulting instrument showed promise as a dictation machine.

Thus challenged, Edison returned to what he later called his favourite invention, and produced in quick succession his 'new', 'improved' and 'perfected' phonographs, of which the last is seen in the illustration. Prominent are an electric cell to drive the motor, three white wax cylinders, and hearing tubes. The *ILN* said of the instrument 'it will be found to be both useful and amusing'.

88 Historic scientific treasures, 1876

An Italian anglophile recently referred to South Kensington as having an 'altissima densità culturale'. This district of London does indeed have a very high cultural density, and we owe this to the Commissioners of the 1851 Exhibition who, strongly influenced by the Prince Consort, devoted the profits of the exhibition to the purchase of the Gore Estate. One institution, set up there 'to extend the influence of science and art on productive industry', was the South Kensington Museum, opened in 1857.

The 'non-art' material assembled in the Museum was of a somewhat miscellaneous character including, for example, sections dealing with food, animal products, building materials and educational apparatus. Three events helped to transform this unpromising mix into a world-famous national museum of science and industry. They were the addition in 1867 of a section on machinery and invention, the acquisition in 1885 of the Patent Office collection, and the display in 1876 of a Special Loan Collection of Scientific Apparatus. The *ILN* illustration shows some of the historical treasures in this Loan Collection.

This was an exhibition, not a trade fair. It was big; it was international; it was opened by the Queen; it focused attention on science as well as engineering as a determinant of social change. It pointed to the need for an independent Science Museum. Many of the exhibited objects were generously presented by their exhibitors. The *ILN* illustration shows some of the historical treasures in this Loan Collection.

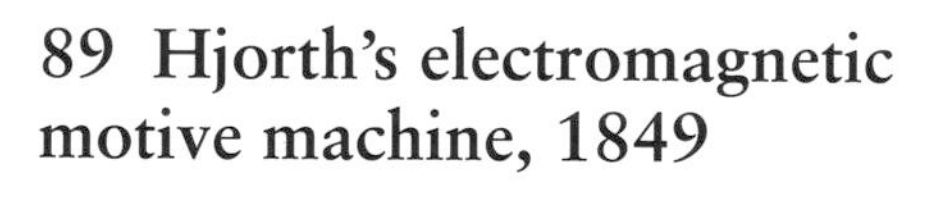

89 Hjorth's electromagnetic motive machine, 1849

Illustrated is the 'electromagnetic engine' (electric motor) which the Dane, Søren Hjorth (1801–70) showed at the Great Exhibition in 1851. The device had no great merit; nevertheless its inventor does deserve a modest mention in the history of electrical technology.

Hjorth realized that the future lay with electric motors driven, not by a wet battery, but by a central generator. This is essentially a coil which rotates in the field of a magnet; as it does, an electrical voltage is induced into it, which can send a current through an external circuit. Hjorth had the brilliant idea that the magnet could be an electromagnet, energized by current generated by the machine itself. He tried to incorporate this principle in a machine but it was not successful. The principle of 'self-excitation' was introduced successfully in 1867 by C. Wheatstone, S.A. Varley and C.W. Siemens working independently.

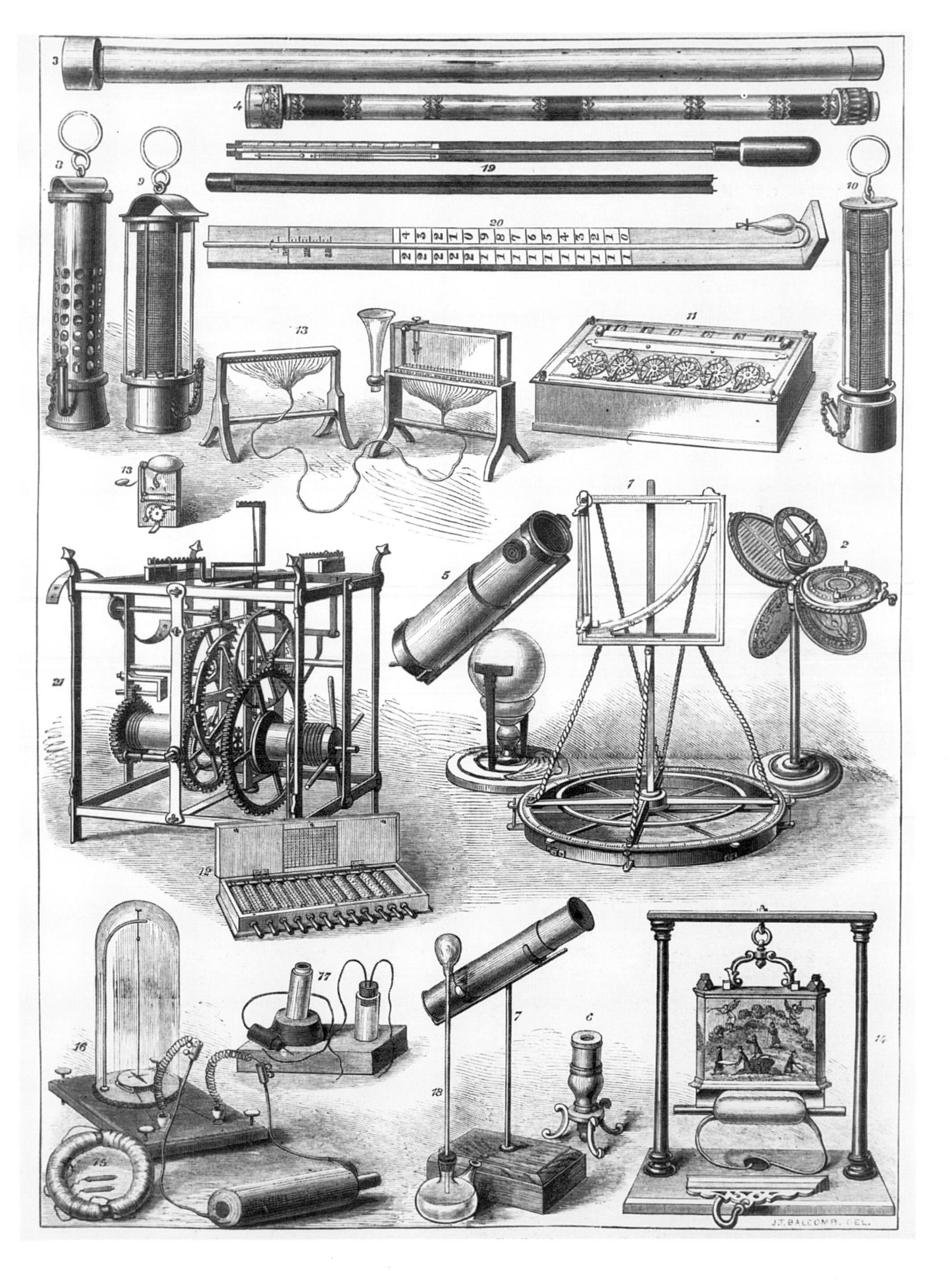
J.T. BALCOMB, DEL.

90 Brunel's 'monstrosity' and Paxton's palace, 1850

Two pictures published by the *ILN* had catalytic effects on the reaction leading to the Great Exhibition, one negative, one positive. On 22 June 1850 it published the plan for the exhibition building which the Building Committee of the Royal Commission had asked I.K. Brunel to prepare, after rejecting every one of the 245 designs submitted in a public competition.

This picture focused all the opposition to the exhibition. There were those who deplored the very idea of an exhibition (all the bad characters at present scattered over the country would be attracted to Hyde Park). There were those who disapproved of the temporary alienation of a part of Hyde Park, and there were those who saw in this too too solid building of sixteen million bricks, which *The Times* called a 'monstrosity', the certainty of permanent alienation.

The second seminal picture was that which Joseph Paxton 'leaked' to the *ILN* (with a readership of the order of 200,000 at the time) and which was published on 6 July 1850. Though he had not seen the official plan, Paxton had heard that it was likely to be unsatisfactory; he had secured a fortnight's grace and, starting from a doodle made on a piece of blotting paper at a meeting, had in nine days produced the plan for the glass and iron building which *Punch* some months afterwards christened the Crystal Palace.

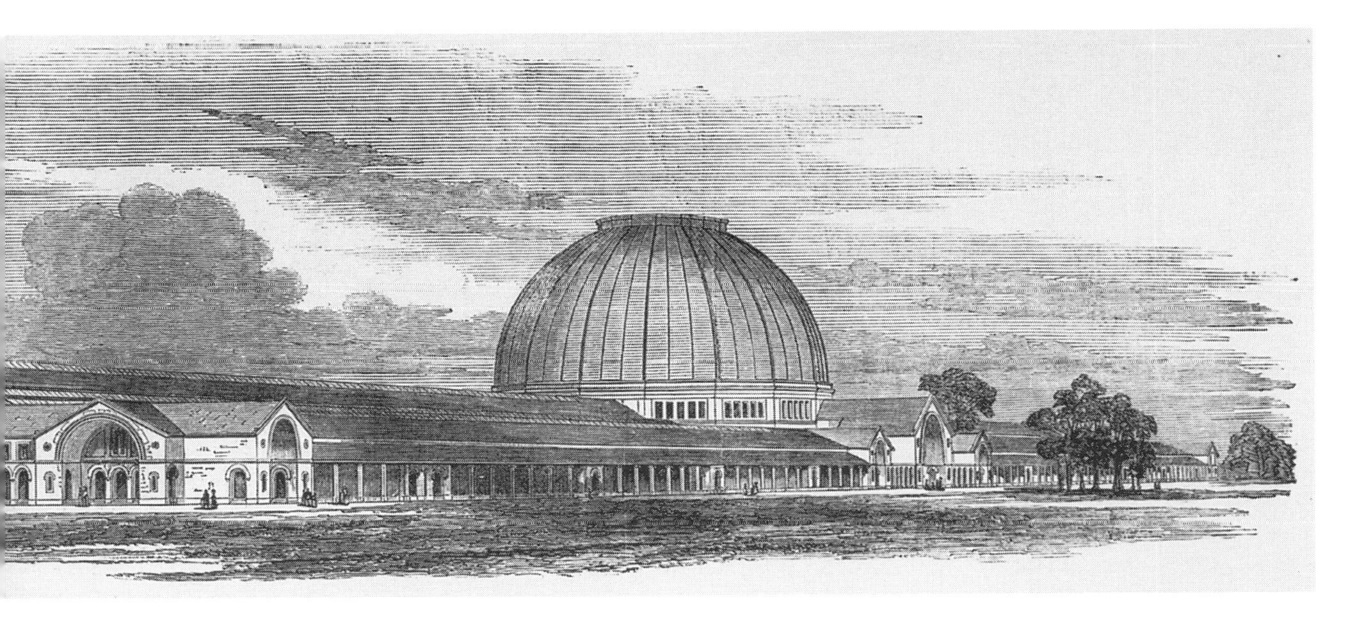

91 The Great Exhibition conceived, 1849

Henry Cole (1808–82), here photographed in old age by Barraud, was educated at Christ's Hospital. When we meet him he was employed in the Public Record Office. His job was no sinecure; in it he was active, reformist and constructive, yet he had enough spare time to become a competent musician and artist, to help in promoting post office reform, to edit a journal of design, to write guide books, to edit *Felix Summerly's Home Treasury*, containing illustrated children's stories, and to publish the first ever Christmas card.

Cole became a member of the Society of Arts at a time when it was feeling its way, through a series of annual exhibitions beginning in 1847, towards a great national exhibition, and he was the most active member of a movement which, through the agency of a Royal Commission presided over by the Prince Consort, brought this about in 1851.

92 The transept under construction, 1850

➸

The bitterest enemy of the Great Exhibition was the notorious Colonel Sibthorp MP whom Gibbs-Smith described as a 'preposterous character'. He regarded the exhibition as 'the greatest trash, the greatest fraud, and the greatest imposition ever attempted to be palmed upon the people of this country'.

Sibthorp opened a Parliamentary attack on 18 June 1850 by asking why a clump of young elm trees was marked out to be cut down on the exhibition site in Hyde Park. These trees were in fact cut down, except one. However, there remained three giant elms standing right across the site. At a lecture to the Society of Arts in November 1850, Paxton (designer of the building) said, 'It may be important here to state that it is unnecessary to cut down any of the large timber-trees, provision being made by means of a curvilinear roof over the transept of the building for them to stand beneath the glass, and by a proper diffusion of air they will not suffer by enclosure.' On a later occasion Paxton said that the idea that there should be a transept at all was due to Henderson, of Fox and Henderson contractors, but the idea that it should have a semi-circular roof was his own.

The *ILN* illustration shows the operation of raising the semi-circular ribs of the transept.

93 Forge used in building the International Exhibition, 1862

The contractor for the 1862 Exhibition was a firm formed by a partnership between Mr (later Sir) John Kelk and the brothers Charles and Thomas Lucas. After apprenticeship with Thomas Cubitt (who 'raised his trade to the dignity of a profession') Kelk practised as a builder for some years and then became a railway contractor, in association with Brassey, Peto and others. After the 1851 Exhibition, which yielded a profit of some £186,000, the Commissioners of that Exhibition, aiming at the purchase of the Gore Estate, sought the aid of a capitalist to treat for it who 'might reasonably be supposed to require it for his own purpose'. Kelk performed this service and so the Commissioners obtained the estate at a much lower price than they would have paid had they bid for it in their own names.

Other works for which Kelk was contractor include the Albert Memorial, Millwall Docks, the Victoria Station and Pimlico Railway, and the Alexandra Palace. The latter was burnt down within a few days of its opening in 1873 and although swiftly rebuilt proved to be a commercial failure.

The forge shown in the illustration was one of several in a large shed stretching along the western side of the exhibition building. 'It is modestly designated "the smiths' shop" but is in fact a cyclopean barrack, in which numerous fires are roaring and glittering, and hundreds of stalwart workmen are making the anvils ring with their massive blows'.

94 Platt's mule in the 1862 Exhibition

Prominent in the International Exhibition was the exhibit of Platt Brothers and Company, a flourishing Lancashire firm specializing in the manufacture of cotton-processing machinery. It consisted of a set of machinery allowing the visitor to follow the processing of cotton through every stage from plucking the pod to spinning finished yarn. The last and largest machine in the sequence was the mule, which as its name implies was a hybrid, combining features of Arkwright's spinning machine and Hargreaves' jenny to emulate what in earlier years had been done by the spinster at her wheel.

Mules were the mainstay of the cotton-spinning industry, to the extent that John Platt MP, senior partner in the firm, could boast in 1866 that the current output of cotton yarn from all the mules in Britain was sufficient to girdle the Earth four times every minute. The example shown was fully automatic. Powered by steam and watched over by one or two 'minders', it did the work of 648 hand-spinners.

95 International Exhibition 1862, machinery in motion

The 1862 Exhibition, like that of 1851, was conceived by the Society of Arts. Originally intended for 1861, it was postponed because of the Franco-Austrian War of 1859. The proposal met with no opposition (Colonel Sibthorp had died in 1855) and confidence in it was attested by a guarantee fund to which more than 1,100 signatories promised about £450,000. The trustees of the fund became Royal Commissioners. Their predecessors of 1851 lent a site south of that part of their Kensington estate which had been leased to the Horticultural Society.

When the exhibition opened, the *ILN* reporter was much moved by the sight of the area devoted to machinery in motion, shown here. 'We do not think there is any point in the whole exhibition the view from which can compare with this in real interest; the array of engines and machines is not only wonderful and beautiful in itself, but every machine exhibited has claims upon our attention as playing some part in our daily wants as luxuries or necessities. By the aid of some of these machines we traverse the oceans of the world and penetrate to the remotest corners of the earth; with others we travel overland at lightning speed and the maximum of comfort; some contribute to the furniture of our houses, others produce the clothes we wear, and many perfect and cheapen the preparation of our daily food. . . .'

THE EDISON DYNAMO MACHINES.

SUBMARINE·MINES

THE SIEMENS CHANDELIER

THE EDISON EXHIBIT

MOUNTAIN TELEGRAPH EQUIPAGE

CHANDELIER OF THE ELECTRIC LIGHT AND POWER GENERATOR COMPANY

THE BRUSH SYSTEM EXHIBIT.

96 The International Electric Exhibition at the Crystal Palace, 1882

This exhibition was promoted by the Crystal Palace Company. Its opening was consistent with a well-known law of nature that 'no international exhibition is ever in readiness at the time first publicly announced'. There was no ceremonial opening; instead on 31 January 1882 the manager 'sent official notices to all exhibitors, stating that the exhibition was to be considered as fairly started from that day'.

Despite this inauspicious beginning the *ILN* reviewed the exhibition favourably. It had fourteen sections, of which one was devoted to 'the refulgence today shed by the influence of a subtle and invisible force' (i.e. to the electric light). The whole exhibition was lit by electic lights supplied by numerous manufacturers. The *ILN* was impressed by the Edison exhibit, including the *Long-waisted Mary Ann* dynamos. However, the object elevated on the altar-like table in the main Edison exhibit was not an electrical instrument at all; it was a tinfoil phonograph (invented 1877).

The *ILN* reporter, viewing the engines of destruction exhibited by the War Office, was 'pervaded by a feeling akin to revulsion'. However, he must have been consoled by seeing the telephone exhibited as enhancing the peaceful art of music; the organ was played in the concert hall and heard by listeners in the telephone room.

97 L'Exposition Universelle, Paris, 1889

The Exposition Universelle of 1889 posed a diplomatic problem because it coincided with the centenary of the French Revolution. Although the French did their best to make the exhibition itself apolitical, European monarchies, painfully reminded that monarchs were decapitable, were not represented officially, and the informal British Empire contribution was organized by a council set up by the Lord Mayor of London.

After the last click of the turnstiles there remained two great monuments. One was a wrought-iron structure, springing from a 100 square metre base to a world record height of 300 metres. This, when projected, had been described by a group of writers, poets, sculptors, architects and amateurs enamoured of beauty as that 'useless and monstrous Eiffel tower'. Eiffel said 'when it is finished they will love it', and love it they did. It was also not entirely useless: it served as 'a beautiful climbing frame for energetic tourists'.

The second monument remained, alas, for only twenty-one years. It was the elegant steel and glass canopy with which Cotamin covered the Galerie des Machines, 'one of the loveliest shapes in which man had ever enclosed space', having nearly twice the width and height of St Pancras Station, London.

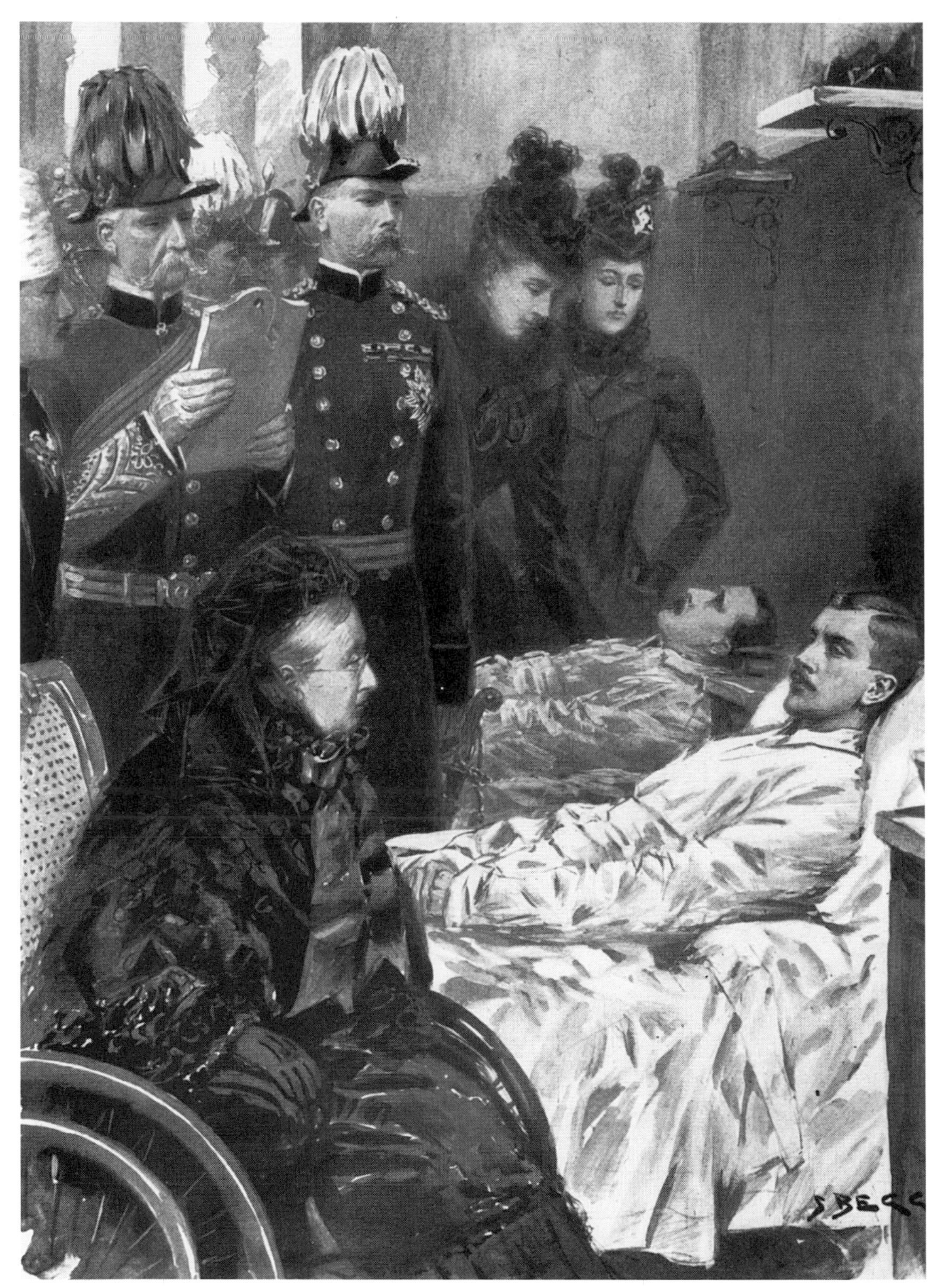

The Queen visits the wounded at Netley Hospital, 1898

War

The remaining illustrations all relate to war. Only one, that of an armoured train, is of land warfare; the major changes in that activity, namely the control of armies by telegraphy, and their transport and supply by strategic railways, did not produce any reproducible illustrations.

Ten of the illustrations, however, refer to naval warfare and here the technology changed very quickly and the changes were well illustrated. We follow these changes by comparing HMS *Caesar* of 1854 with HMS *Hood* of 1893, *Caesar* being one of the earliest warships designed with steam as well as sail propulsion.

Four illustrations show the manufacture and testing of big guns and their firing at the bombardment of Alexandria. Two others are related in that they both refer to a critical period in naval development when the armouring of warships was being discussed and it was realized that a single armoured vessel might defeat a fleet. Finally we show an unusual Russian ironclad of 1875 affectionately known to sailormen as the 'Popoffski'.

98 Rolling armour plate for Her Majesty's ships, 1863

Sir John Brown (1816–96) was born in Sheffield. After education to the age of fourteen and apprenticeship to a merchant, he entered the steel industry in a small way, making files and also conical buffer springs of his own invention. The business prospered and by 1860 he had become a pioneer manufacturer of steel by the Bessemer process.

In 1860 Brown happened to be at Toulon where he saw the French warship *La Gloire,* the first ever armoured line-of-battle ship, the building of which had caused alarm in the United Kingdom. Denied access to her, he deduced by inspection at a distance the dimensions of her plates, and decided that they were hammered rather than rolled. He was confident that bigger, thicker and stronger plates could be made by rolling, and on return he established at his Atlas Works a rolling mill to produce armour plate, initially 4½ inches thick.

Brown described the last stage of manufacture, illustrated here, in these terms: 'to drag the pile out of the furnace, convey it to the rolls and force it between them in so short a time to avoid its losing the welding heat is a matter of greater difficulty than those unacquainted with the work would imagine. The intensity of the heat thrown off is almost unendurable, and the loss of a few moments in the conveyance . . . from the furnace to the rolls is fatal to the success of the operation.'

99 Measuring the speed of a projectile, 1864

The projectile broke a first wire, thereby starting electrically a chronoscope in the hut, and then it broke a second wire at a known distance from the first, thereby stopping the chronoscope. The instrument was essentially a pendulum; one could deduce a time interval from the angle through which it swung between the start and stop signals.

This chronoscope was invented by Navez in Belgium. On it the following judgement was passed by the Revd Francis Bashforth: 'perhaps no instrument has had so much money wasted upon it in the way of experiment as that of Navez, and yet it gave no results of any value'. But Bashforth was not disinterested. He combined the duty of Rector of Minting near Horncastle with that of Professor of Applied Mathematics to the Advanced Class of the Royal Artillery Officers, Woolwich, and in the latter capacity invented his own chronoscope. It is now in the collections of the Science Museum.

100 Paris 1870: departure of a balloon by night

On 19 July 1870 France declared war on Prussia. By 2 September Marshal MacMahon had been defeated at Sedan, the Emperor of the French was a captive, and Marshal Bazaine and his army were besieged in Metz. Prussian armies moved to surround Paris, and the only communication between the capital and unoccupied France was to be by air. So two great railway termini, the Gare d'Orléans and the Gare du Nord, became factories, where women stitched calico balloons of some 1,980 cubic metre capacity, and sailors rigged them. Sailors also piloted some of them on their perilous missions.

Ballooning was a one-way traffic: passengers and mail were carried out of the city, but no balloon made its way back. The uncertainties of the journey are exemplified by the adventure of the balloon *Ville d'Orléans* which, setting out for Tours where the provisional government of the Republic had been set up, ended up in Norway. The most important living creatures carried by a balloon were not the passengers but the homing pigeons. Of the birds that were conscripted during the war some two hundred were taken out of the city by balloon to carry the inward post. Only about sixty completed their missions, the others falling victim to fog, snow and (it is said) hawks imported from Germany. Nevertheless an enormous amount of information was carried by these avian heroes owing to the brilliant technological achievement of the photographer Dagron whose process of microphotography enabled one pigeon to carry up to forty thousand despatches.

101 The Tsar inspects the 'infant school' at Woolwich, 1874

In 1874 Tsar Alexander II of Russia paid a visit to England. Soon after the imperial yacht left Flushing it ran aground. On refloating, to make up for lost time, it headed for Dover instead of Gravesend where the red carpet was down. However the improvised reception went well and His Majesty departed by rail for Windsor via London Bridge and Waterloo. The subsequent programme included visits to the Crystal Palace (at Sydenham), the City of London and the Albert Hall, a review of troops at Aldershot and an artillery review at Woolwich, preceded by an inspection of the arsenal.

The position as regards naval gunnery at this time was that breech loading, introduced in 1861, had been abandoned in 1867, and a race between guns and armour was being carried on using muzzle loaders. Of those mounted on *Warrior* (Britain's first ironclad), when she was rearmed in 1867, the largest were of 6-inch calibre and weighed 9 tons. Seven years later the Tsar saw at Woolwich a gun of the type mounted on *Devastation* (1873) having 12-inch calibre and weighing 35 tons. According to the *ILN* this monster had been called the Woolwich Infant 'by way of a joke on its size'. The Tsar saw also some fifty other guns, large and small, 'which some facetious officer has called the infant school'.

102 Firing the 80-ton gun, 1877

The era of the monster naval gun was ushered in, not by the French or the British, but by the Italians who, shortly after the unification of their country in 1870, began a spirited programme of naval construction under the gifted designer Benedetto Brin (1833–98). The first fruits of this were *Duilio* (10,650 tons) and her sister ship *Dandolo*.

Originally these ships were designed to be armed with four 35-ton guns, but when Armstrong of Elswick (103) declared himself capable of making a bigger gun, this was changed to 60 tons. At this time the British were designing the battleship *Inflexible*, intended to carry 60-ton guns. This was now changed to 80 tons. In the illustration it is a muzzle-loading gun of this size, made at Woolwich, which is shown under test.

Meanwhile the Italians had gone one better and mounted in *Duilio* not 60-ton guns but Armstrong's 100-ton 17.7-inch muzzle loaders. The *ILN* complained that 'we have four 80-ton guns afloat against the eight Italian 100-ton guns' and it pointed out that the Italian guns were 50 per cent more powerful than the British.

103 Making a gun barrel by Armstrong's method, 1883

William Armstrong (1810–1900), armaments manufacturer, was born in Newcastle. A solicitor for some years, he found greater interest in engineering than in law and eventually (1847) established at Elswick a factory manufacturing hydraulic machinery.

The reported inadequacy of British guns in the Crimea attached his attention. His design for a rifled, breech-loaded gun was adopted for service in the field, and in 1859 he was appointed Engineer of Rifled Ordnance to the War Department. He thereupon set up at Elswick a company to make and sell to the government the weapons designed at Woolwich.

Here he developed a method of making gun barrels whereby several wrought-iron cylinders were shrunk onto a steel core. Each cylinder was made by winding a white-hot iron bar round a steel former and then welding adjacent coils together by blows from a steam hammer. The dramatic *ILN* picture shows this procedure in operation at Shanghai Arsenal.

104 HMS *Caesar*, 1854

The Royal Navy did not welcome steam propulsion with open arms. This was not entirely due to conservatism: the unreliability of early steam engines, the vulnerability of paddles, the fact that paddles denied space to guns, and the absence of overseas coaling stations, all tended to restrict the use of unassisted steam to small craft such as tugs and despatch vessels; in larger vessels it had to be auxiliary to sail, and it had no place at all in the line of battle.

The situation changed with the advent of the screw propeller, whose superiority to the paddle was confirmed by a tug of war in 1845 between the screw ship *Rattler* and her sister with paddles, *Alecto*. In 1846 auxiliary screw propulsion was added to the existing line-of-battle ship *Ajax*, and in 1852 *Agamemnon* was launched, the first ship of the line to be designed as a screw steamer.

Caesar, of 1854, resembled *Agamemnon*. She is here seen in November being hastily fitted out to take part in the Crimean War which had broken out in March. Her armament consisted of seventy-four 32-pounders firing shot and sixteen 8-inch guns firing shells. Thus *Caesar*, like Nelson's ships at Trafalgar, was a 'wooden wall' and she fired broadsides. However, she had made two steps in evolution: she could fire explosive shells and she had auxiliary steam.

105 *Merrimack* rams *Cumberland*, 1862

'Is it indeed true that the naval supremacy of England has passed away like a mere unsubstantial exhalation . . . ?' asked the *ILN* on 5 April 1862. The reason for the alarm was this. At the beginning of the American Civil War the Federals, obliged to leave Norfolk to the Confederates, partly burned and then sank their first-class frigate *Merrimack*. The Confederates raised her and built on her deck a structure of thick armour with flat roof and sloping sides, designed to be proof against shot and shell. Inside it were mounted six 9-inch, two 7-inch and two 6-inch guns. Renamed *Virginia* this strange ship met the Federal fleet in Hampton Roads, rammed and sank the sailing frigate *Cumberland* (see illustration) and forced the surrender of the *Congress*.

On the next day *Virginia* found herself up against, not the anticipated third victim *Minnesota* but John Ericsson's *Monitor*, which was essentially a steam-operated rotating turret, made of iron 8 inches thick and housing two 11-inch Dahlgren guns, mounted on an iron platform which projected beyond a wooden hull. The two ships fought each other for four hours but at the end neither had gained the advantage. Armour had made them vulnerable to existing gunnery: turret mounting had enabled two heavy guns to hold their own against ten lighter ones. According to *The Times*, before the action the United Kingdom had 149 first-class warships; now she had only two – her two ironclads, *Warrior* and *Black Prince*.

106 *Novgorod*, circular ironclad, 1875

Novgorod was one of two sister ships designed by Vice-Admiral Popov for the Russian Navy, the second bearing his own name. The ship was of 101 feet diameter. She carried two 11-inch guns, mounted in a barbette at the centre. The guns were fixed in direction and were aimed by rotating the ship. There were six screws driven by eight steam engines, rotation being achieved by the use of the outer pair. The freeboard was about 18 inches.

There was method in the apparent madness of the revolutionary design. The object of the circularity was to provide a stable platform for the guns, one which would neither pitch nor roll, and the object of the low draught was to enable the ship to carry out its duties of coastal defence in shallow water.

107 Diving, 1878

On 24 March 1878 the frigate HMS *Eurydice* (921 tons), a training ship for young ordinary seamen, was returning from a cruise in the West Indies with 366 on board, about 300 being men under training and the remainder military personnel. Sensing deterioration of the weather the captain was beginning to shorten sail, when there was a sudden change of wind from westward to eastward, and shortly afterwards the ship capsized and sank at a point about three miles from Dunnose, between Shanklin and Ventnor, Isle of Wight. Five survivors were rescued by the schooner *Emma*; three of them died soon afterwards.

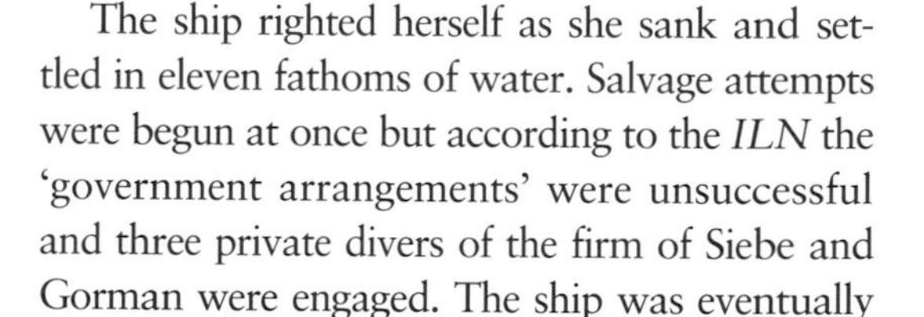

The ship righted herself as she sank and settled in eleven fathoms of water. Salvage attempts were begun at once but according to the *ILN* the 'government arrangements' were unsuccessful and three private divers of the firm of Siebe and Gorman were engaged. The ship was eventually towed into Portsmouth on 1 September 1878.

The interesting feature of the illustration is the electric lamp used by the diver. It is an arc lamp with automatic regulation, of a type devised by Foucault, and the case is by Siebe and Gorman. The battery is of fifty Bunsen cells, in boxes of ten. The arc was said to last for one hour in the open air and two in the case.

108 HMS *Alexandra* in the bombardment of Alexandria, 1882

In 1880 Egypt was nominally a vassal state of Turkey, ruled locally by the Khedive. But in order to protect the interests of European creditors, chiefly British and French, financial control over the near-bankrupt government was exercised by Sir Evelyn Baring and M. de Blignières. A revolt against both Turkish and European influences was led by one Arabi Pasha, and anti-European riots occurred in Alexandria in 1882. An international fleet assembled there, and when the Egyptian army leaders defied an order to stop improving the defences of the city, the fortifications were on 11 July 1882 bombarded by the British element of the fleet (the French being then opposed to military intervention).

The British commander of the operation was Admiral Sir Frederick Beauchamp Seymour. He had at his disposal eight battleships and five gunboats. One of the battleships was the *Alexandra* (1877, 325 ft, 9,490 tons), which was a perfect example of the mounting of all the main armament in a central battery or citadel. This armament consisted (originally) of two 11-inch and ten 10-inch muzzle loaders on two decks, with six guns providing conventional broadside fire and six capable of fore-and-aft fire. In the bombardment *Alexandra* fired forty-eight 11-inch and 221 10-inch shells.

109 Armoured train, 1882

The circumstances leading up to the bombardment of Alexandria by the Royal Navy in 1882 have been described in 108. Among the many who regarded this action as futile was Lieutenant General Sir Garnet Wolseley (1833–1913), later Field Marshal and Viscount, who commanded the military force that was sent to Egypt to conduct land operations against Arabi Pasha. Wolseley left England on 15 August 1882 and by a brief and brilliant campaign, culminating in a night attack on 13 September, he routed Arabi's army.

A device that played a useful part in the early phase of this campaign was the armoured train illustrated, manned by seamen of the Naval Brigade. It consisted of five trucks and a locomotive. In the foremost truck was a Nordenfeldt gun, and the men were protected by iron shields. The locomotive was sandbagged. In the rearmost truck were three Gatling guns and it was proposed to add a 40-pounder gun. The last truck but one carried a steam crane.

Interest is added to this illustration by the fact that the train was fitted out and commanded by a man who was then a Captain RN but who ultimately became Admiral of the Fleet, Baron Fisher of Kilverstone (1841–1920), 'one of the greatest administrators in the history of the Royal Navy'.

110 HMS *Hood*, 1893

A portrait of HMS *Hood* shows the very great change in naval technology that took place in the half-century following the commissioning of HMS *Caesar* (104). *Caesar* is 207 feet long, *Hood* 380 feet. *Caesar* displaces 2,800 tons, *Hood* 14,150 tons. *Caesar* is of wood, while *Hood* is of steel, a development made possible by the Siemens-Martin process for steel manufacture. *Caesar* is unarmoured; *Hood* is armoured at vital areas, this armour including a belt of steel up to 18 inches thick. *Caesar* is fully rigged and its rigging almost conceals its apologetic-looking funnel; *Hood* has two masts but they have no propulsion function.

The differences are greatest as regards armament. *Caesar* has ninety cast-iron guns, seventy-four firing shot and sixteen shell, all smooth bore and muzzle loading. *Hood*'s armament is sharply divided into primary and secondary, the primary being four 13.5-inch guns, rifled and breech loading. *Caesar*'s armament is arranged on three decks so as to fire broadsides. *Hood*'s big guns are in pairs in rotating turrets, each with a very wide angle of fire.

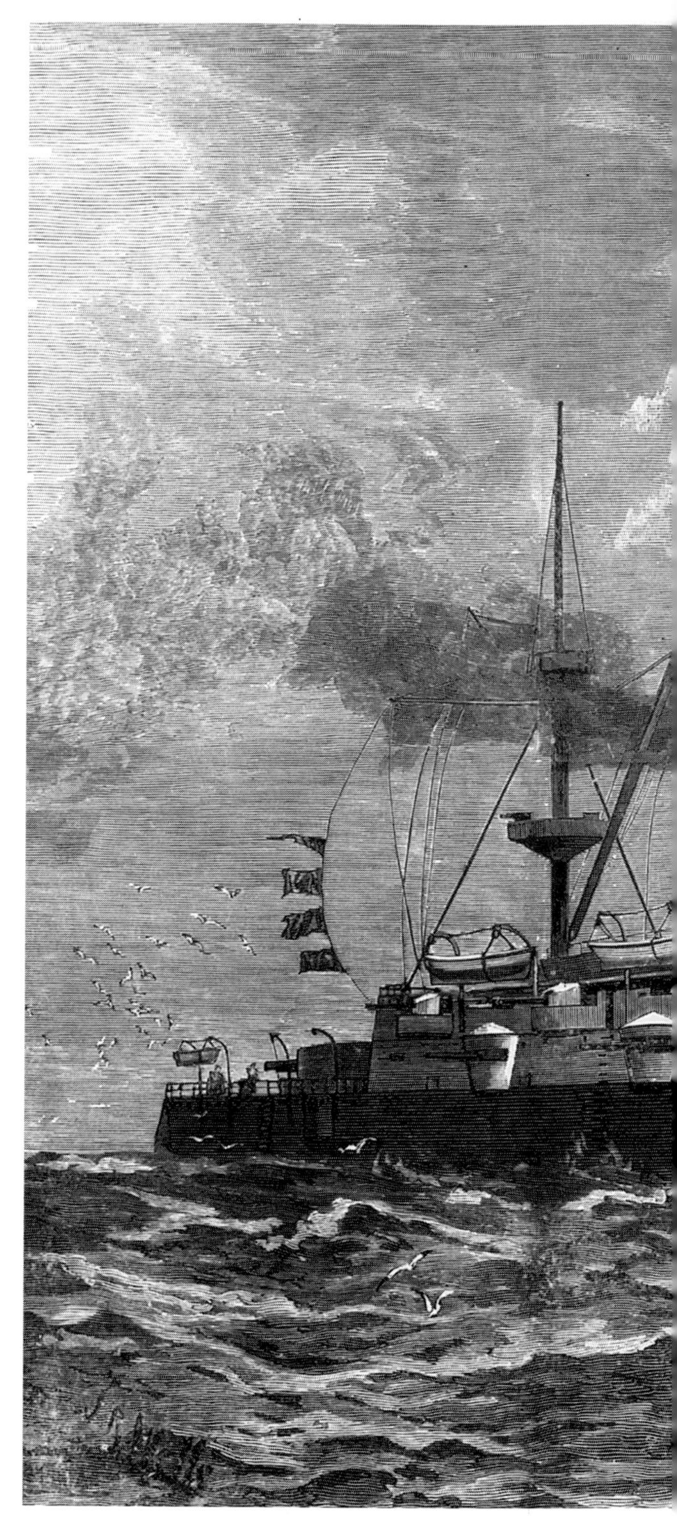

Sources

An *ILN* source is given for each picture. Dimensions are given in centimetres, or as 'wp' (whole page *c.* 36 × 24 cm), or 'dp' (double page, *c.* 51 × 36 cm). Where one or more books or journals were of particular value in preparing the text, these are cited; details can be found in the bibliography. Except where otherwise attributed, quotations in the text are from the *ILN* article which accompanied the picture.

Frontis. Vol. 54, p. 401, 1869, 24 × 17 (detail)

Introduction

p. x Vol. 37, p. 306, 1860, 15 × 20
p. xi Vol. 75, p. 209, 1879, 22 × 15

Food, Work and Health

Frontis. Vol. 68, frontispiece, 1876, foldout, 48 × 33
1 Vol. 31, p. 165, 1857, wp; Rolt (1974), Williams (1977)
2 Vol. 41, p. 652, 1862, wp; Trow-Smith (1980)
3 Vol. 71, p. 544, 1877, wp; Hobhouse (1985)
4 Vol. 93, p. 641, 1888, wp; Vaizey (1960)
5 Vol. 91, pp. 84–5, 1887, dp; Pevsner (1992)
6 Vol. 51, p. 421, 1867, wp; Marshall (1981), Carr and Taplin (1962)
7 Vol. 60, p. 4, 1872, wp; Metcalfe (1969), Robinson (1951)
8 Vol. 63, p. 504, 1873, wp; Smith (1961)
9 Vol. 51, p. 565, 1867, wp; Anderson (1982)
10 Vol. 18, p. 149, 1851, 24 × 15; Whalley (1975)
11 Vol. 45, p. 393, 1864, 24 × 16; Warren (1970)
12 Vol. 65, p. 460, 1874, 24 × 25; Timmins (1866)
13 Vol. 67, p. 197, 1875, 24 × 17; Bessemer (1989)
14 Vol. 83, p. 181, 1883, wp; Neff (1966)
15 Vol. 89, p. 529, 1886, wp; Wilson (1954)
16 Vol. 33, p. 510, 1858, 24 × 17; Everard (1949)
17 Vol. 96, p. 164, 1890, 23 × 17
18 Vol. 50, p. 632, 1867, wp
19 Vol. 94, p. 244, 1889, wp; Rolt (1974)
20 Vol. 41, p. 13, 1862, 22 × 16; Davies (1989)
21 Vol. 53, p. 161, 1868, 24 × 17; Institution of Civil Engineers (1864/5)
22 Vol. 52, p. 264, 1868, wp; Gilbert (1969)
23 Vol. 26, p. 176, 1855, 24 × 15
24 Vol. 101, p. 388, 1892, wp
25 Vol. 97, p. 688, 1890, 24 × 28; Smith (1980)
26 Vol. 84, p. 592, 1884, 7 × 11; Nicolle (1961)
27 Vol. 110, p. 37, 1897, wp

Communications and Transport

Frontis. Vol. 104, p. 637, 1894, wp
28 Vol. 6, p. 233, 1845, 24 × 15; Marland (1964)
29 Vol. 21, p. 205, 1852, 16 × 22; de Carle (1947)
30 Vol. 31, p. 108, 1857, wp; Dibner (1964)
31 Vol. 46, p. 101, 1865, 24 × 23
32 Vol. 49, p. 357, 1866, wp
33 Vol. 80, p. 10, 1882, 22 × 15; de Sola Pool (1977)
34 Vol. 3, p. 73, 1843, 23 × 16
35 Vol. 9, p. 212, 1846, 25 × 13; Corlett (1990)
36 Vol. 31, p.489, 1857, wp; Dugan (1953), Beaver (1969)
37 Vol. 35, p. 303, 1859, wp; Crook (1990)
38 Vol. 36, pp. 600–1, 1860, dp; Dugan (1953), Caldwell (1976)
39 Vol. 84, p. 9, 1884, wp; Hague and Christie (1975)

40 Vol. 8, p. 369, 1846, wp; Day (1985)
41 Vol. 6, p. 137, 1845, 24 × 16; Clayton (1966)
42 Vol. 45, p. 276, 1864, 24 × 18; Clayton (1966)
43 Vol. 12, p. 170, 1848, 24 × 15; Rolt (1960)
44 Vol. 16, p. 192, 1850, wp
45 Vol. 19, p. 604, 1851, 24 × 13
46 Vol. 21, p. 552, 1852, 24 × 15; Livesey (1948)
47 Vol. 16, p. 433, 1850, 24 × 13; Day (1960)
48 Vol. 34, p. 533, 1859, 24 × 16; Bowden (1983)
49 Vol. 46, p. 372, 1865, wp; Course (1962)
50 Vol. 67, p. 149, 1875, wp; Government of India (1953)
51 Vol. 49, p. 232, 1866, wp; Jackson (1985)
52 Vol. 40, pp. 64–5, 1862, dp; Greenhill (1984)
53 Vol. 59, p. 228, 1871, wp; Simmons (1968)
54 Vol. 76, p. 257, 1880, wp; Beaver (1972)
55 Vol. 75, p. 9, 1879, wp; Institution of Civil Engineers (1881)
56 Vol. 76, pp. 20–1, 1880, dp; Engineer (1880), Prebble (1975), Thomas (1972)
57 Vol. 41, p. 692, 1862, wp; Menear (1983)
58 Vol. 52, p. 141, 1868, 17 × 20; Day (1979)
59 Vol. 95, pp. 500–1, 1889, dp; Westhofen (1890)
60 Vol. 75, p. 481, 1879, 22 × 16; Behrend (1962)
61 Vol. 101, p. 601, 1892, 23 × 28; Rolt (1976)
62 Vol. 36, p. 133, 1860, 24 × 15; Kidner (1950)
63 Vol. 45, p. 605, 1864, wp; Body (1976)
64 Vol. 50, p. 596, 1867, wp; Goode (1869)
65 Vol. 82, p. 276, 1883, 23 × 16
66 Vol. 54, p. 256, 1869, 24 × 15; McGurn (1987)
67 Vol. 84, p. 577, 1884, wp
68 Vol. 49, p. 581, 1866, 24 × 17; Whitehead (1975)
69 Vol. 115, Christmas Number, p. 11, 1899, wp

Science: Created, Explained, Applied and Exhibited

Frontis. Vol. 18, p. 349, 1851, dp
70 Vol. 38, p. 28, 1861, 24 × 24; Bowers (1991)
71 Vol. 58, p. 244, 1871, 24 × 30; Darwin (1887)
72 Vol. 57, p. 296, 1870, wp; Huxley (1903)
73 Vol. 101, facing p. 804 (supplement), 1892, wp
74 Vol. 42, p. 481, 1863, 24 × 17; Pepper (1890), Dircks (1863)
75 Vol. 31, p. 5, 1857, 24 × 18
76 Vol. 56, p. 493, 1870, 24 × 23; Thompson (1981)
77 Vol. 22, p. 253, 1853, 16 × 16
78 Upper: Vol. 29, p. 639, 1856, 24 × 19; lower: Vol. 32, p. 225, 1858, 24 × 17; Darwin (1986)
79 Vol. 90, p. 197, 1887, wp
80 Vol. 76, p. 101, 1880, wp
81 Vol. 52, p. 5, 1868, wp; Airy (1896), McCrea (1975)
82 Vol. 77, p. 565, 1880, wp; Meadows (1975)
83 Vol. 78, p. 364, 1881, wp; Royal Society (1932)
84 Vol. 47, p. 256, 1865, 24 × 16; Gernsheim (1969)
85 Vol. 94, p. 645, 1889, wp; Haas (1976)
86 Vol. 109, facing p. 66, 1896, wp; Nitske (1971)
87 Vol. 93, p. 81, 1888, wp; Chew (1981)
88 Vol. 69, p. 269, 1876, wp
89 Vol. 14, p. 309, 1849, 17 × 19
90 Upper: Vol. 16, p. 445, 1850, 35 × 9; lower: Vol. 17, p. 13, 1850, 35 × 11
91 Vol. 63, p. 36, 1873, wp; *Dictionary of National Biography*
92 Vol. 17, p. 453, 1850, wp; Gibbs-Smith (1981)
93 Vol. 40, p. 163, 1862, wp; Institution of Civil Engineers (1886/7)
94 Vol. 41, p. 168, 1862, wp; Platt (1866)
95 Vol. 41, pp. 420–1, 1862, dp
96 Vol. 80, p. 204, 1882, wp
97 Vol. 95, following p. 586 (supplement), 1889, dp; Allwood (1977)

War

Frontis. Vol. 113, p. 873, 1898, wp
98 Vol. 39, p. 274, 1861, wp; *100 Years in Steel* (1937)
99 Vol. 45, p. 485, 1864, 24 × 18; Bashforth (1873)
100 Vol. 57, p. 676, 1870, wp; Fisher (1965)
101 Vol. 64, p. 601, 1874, wp
102 Vol. 70, facing p. 144, 1877, foldout, 45 × 31
103 Vol. 83, pp. 100–1 (supplement), 1883, dp; McKenzie (1983)
104 Vol. 25, p. 520, 1854, 25 × 29
105 Vol. 40, p. 327, 1862, 24 × 20; Hartman (1983)
106 Vol. 68, p. 24, 1876, foldout, 47 × 31; Kemp (1983)
107 Vol. 72, p. 389, 1878, 24 × 26
108 Vol. 81, pp. 126–7, 1882, dp
109 Vol. 81, pp. 182–3, 1882, dp
110 Vol. 99, p. 165, 1891, wp

Bibliography

Airy, W. (ed.), *Autobiography of Sir George Biddell Airy*, Cambridge University Press, 1896
Allwood, J., *The Great Exhibition*, Studio Vista, 1977
Anderson, D., *Coal*, David and Charles, 1982
Bashforth, F., *The Motion of Projectiles*, Asher and Co., 1873
Beaver, P., *The Big Ship*, Evelyn, 1969
Beaver, P., *A History of Tunnels*, Peter Davies, 1972
Behrend, G., *Pullman in Europe*, Ian Allan, 1962
Body, G., *Clifton Suspension Bridge*, Moonraker Press, 1976
Bowden, T.N., *Brunel's Royal Albert Bridge Saltash*, Peter Watts, 1983
Bowers, B., *Michael Faraday and the Modern World*, EPA Press, 1991
Caldwell, J.B., 'The Three Great Ships' in A. Pugsley (ed.), *The Works of Isambard Kingdom Brunel*, Institution of Civil Engineers, 1976
Carr, J.C. and Taplin, W., *History of the British Steel Industry*, Basil Blackwell, 1962
Chew, V.K., *Talking Machines*, HMSO/Science Museum, 1981
Clayton, H., *The Atmospheric Railways*, published by the author, 1966
Corlett, E., *The Iron Ship*, Conway Maritime Press, 1990
Course, E., *London Railways*, Batsford, 1962
Crook, A.W., 'The Brunels', *Mining Technology*, vol. 72 (1990)
Darwin, F. (ed.), *The Life and Letters of Charles Darwin*, vol. 3, John Murray, 1887
Darwin, J., *The Triumphs of Big Ben*, Robert Hale, 1986
Davies, P., *Troughs and Drinking Fountains*, Chatto and Windus, 1989
Day, J.R., *More Unusual Railways*, Frederick Muller, 1960
Day, J.R., *The Story of London's Underground*, London Transport, 1979
Day, L.R., *Broad Gauge*, HMSO/Science Museum, 1985
de Carle, E., *British Time*, Crosby Lockwood, 1947
de Maré, E.S., *The Victorian Woodblock Illustrators*, Gordon Fraser, 1980
de Sola Pool, I. (ed.), *The Social Impact of the Telephone*, MIT Press, 1977
Dibner, B., *The Atlantic Cable*, Blaisdell Publishing Co., 1964
Dircks, H., *The Ghost*, printed in Greenwich, 1863
Dictionary of National Biography, article 'Cole, Sir Henry'
Dugan, J., *The Great Iron Ship*, Hamish Hamilton, 1953
The Engineer, vol. 50, various entries in 1880
Everard, S., *The History of the Gas Light and Coke Company*, Ernest Benn, 1949

Fisher, J.S., *Airlift* 1870, Max Parrish, 1965
Gernsheim, H., *The History of Photography*, Thames and Hudson, 1969
Gibbs-Smith, C.H., *The Great Exhibition of 1851*, HMSO, 1981
Gilbert, K.R., *Fire Fighting Appliances*, HMSO/Science Museum, 1969
Goode, T., *The Holborn Viaduct*, broadsheet, 1869
Government of India Ministry of Railways, *Indian Railways: One Hundred Years*, 1953
Greenhill, R., *Spanning Niagara*, University of Washington Press, *c.* 1984
Haas, R.B., *Muybridge, Man in Motion*, University of California Press, 1976
Hague, D.B. and Christie, R., *Lighthouses*, Gomer Press, 1975
Hartman, T., *The Guinness Book of Ships and Shipping*, Guinness Superlatives, 1983
Hobhouse, H., *Seeds of Change*, Sidgwick and Jackson, 1985
100 Years in Steel, Thomas Firth and John Brown, 1937
Huxley, L., *Life and Letters of Thomas Henry Huxley*, vol. 2, Macmillan, 1903
Institute of Metals, *Sir Henry Bessemer, FRS – an Autobiography*, 1989
Minutes of Proceedings of the Institution of Civil Engineers, vol. 24 (1864/5)
Minutes of Proceedings of the Institution of Civil Engineers, vol. 63 (1881)
Minutes of Proceedings of the Institution of Civil Engineers, vol. 87 (1886/7)
Jackson, A.A., *London's Termini*, David and Charles, 1985
Kemp, P., *The History of Ships*, Orbis, 1983
Kidner, R.W., *The First Hundred Road Motors*, Oakwood Press, 1950
Livesey, H.F.F., *The Locomotives of the LNWR*, Railway Publishing Co., 1948
Marland, E.A., *Early Electrical Communication*, Abelard-Schuman, 1964
Marshall, J.D., *Furness and the Industrial Revolution*, Michael Moon, 1981
McCrea, W.H., *Royal Greenwich Observatory*, HMSO, 1975
McGurn, J., *On your Bicycle*, John Murray, 1987
McKenzie, P., *W.G. Armstrong*, Longhirst Press, 1983
Meadows, A.J., *Greenwich Observatory*, vol. 2, Taylor and Francis, 1975
Menear, L., *London's Underground Stations*, Midas Books, 1983
Metcalfe, J.E., *British Mining Fields*, Institute of Mining and Metallurgy, 1969
Neff, W.F., *Victorian Working Women*, Frank Cass, 1966
Nicolle, J., *Louis Pasteur*, Hutchinson, 1961
Nitske, W.R., *The Life of Wilhelm Conrad Röntgen*, University of Arizona Press, 1971
Pepper, J.H., *The True History of the Ghost*, Cassell, 1890
Pevsner, N.B.L., *The Buildings of England, Northumberland*, 2nd edn, Penguin, 1992
Platt, J., 'On Machinery for the Preparing and Spinning of Cotton', *Proceedings of the Institution of Mechanical Engineers*, vol. 30 (1866)
Prebble, J., *The High Girders*, Secker and Warburg, 1975
Robinson, K.G., *Wilkie Collins*, The Bodley Head, 1951
Rolt, L.T.C., *George and Robert Stephenson*, Longmans, 1960
Rolt, L.T.C., *Red for Danger*, David and Charles, 1976
Rolt, L.T.C., *Victorian Engineering*, Penguin, 1974
Proceedings of the Royal Society of London, vol. 135 (1932), Obituary Notices
Simmons, J., *St Pancras Station*, Allen and Unwin, 1968
Smith, R., *Sea-coal for London*, Longmans, 1961
Smith, F.B., *The People's Health 1830–1910*, Croom Helm, 1980
Thomas, J., *The Tay Bridge Disaster*, David and Charles, 1972

Thompson, D., in W.H. Brock et al. (eds), *John Tyndall, Essays on a Natural Philosopher,* Royal Dublin Society, 1981

Timmins, S. (ed.), *The Resources, Products and Industrial History of Birmingham and the Midland Hardware District*, printed in Birmingham, 1866

Trow-Smith, R., *History of the Royal Smithfield Club*, Royal Smithfield Club, 1980

Vaizey, J., *The Brewing Industry, 1886–1951*, Pitman, 1960

Warren, J.G.H., *A Century of Locomotive Building by Robert Stephenson and Co.*, David and Charles, 1970

Westhofen, W., 'The Forth Bridge', *Engineering*, vol. 49 (1890)

Whalley, J.I., *Writing Implements and Accessories*, David and Charles, 1975

Whitehead, R., *A Century of Steam-rolling*, Ian Allen, 1975

Williams, M., *Steam Power in Agriculture*, Blandford, 1977

Wilson, C., *The History of Unilever*, vol. 1, Cassell, 1954

Index

Abbey Mills, 26
accidents, 12, 78
Agamemnon, HMS, 38–9
Agricultural Hall, Islington, 3
agriculture, 2–3
air walker, 98
Airy, Sir George, 37, 100, 102
Alexandra, HMS, 138
Allsop, Samuel, 4
Argand, Ami, 50
Armstrong, Sir William, 6, 132–3
Atlantic cable, 38–40

ballooning, 129
Banks, Sir Joseph, 20
Barrow-in-Furness, 8
Barry, Sir Charles, 98
Bashforth, Revd Francis, 128
Bazalgette, Sir Joseph, 22, 26
Belville, J.H., 37
Bessemer, Sir Henry, 17; steel process, 8, 16–17
Bewick, Thomas, xi
bicycle, 86–7
Big Ben, 98
Birmingham, 14, 16
Botallack Mine, 10
Bouch, Sir Thomas, 71, 76
Boydell, James, 2
brewing, 4
bridges: Britannia, 55; Clifton, 82; Conway, 54; Forth, 76; Hungerford, 47, 81; Menai, 54–5; Newcastle, high level and swing, 6; Niagara, 66; Royal Albert, Saltash, 60, 82; Tay, 71–2
Bristol, 42
British Association, 17, 94–5, 105
British Museum, 21, 94
broad gauge, 51
Brown, Sir John, 126
Brunel, Isambard K., 39, 42–4, 47, 51, 54, 60, 82, 112
Burdett Coutts, Angela, 25

Caesar, HMS, 134, 140
Camden Town railway, 56
Cammell, Messrs, 17
Cannon Street Station, 64
car, motor, 89
carriage, steam, 80
cholera, 31–2
chronoscope, 128
Claxton, Captain, 42, 54
Clegg, Samuel, 20
Cole, Henry, 114
Collins, Wilkie, 10
communications, 36–41
Cooke, William, 36
Cory, Wm and Son, 11
Crystal Palace: Hyde Park, 112; Sydenham, 53, 121
Cumberland, 135

Dam, Vyrnwy, 24
Darwin, Charles, 93
Denison, Edmund, 98
Dircks, Henry, 95
diving, 136
drinking fountain, 25
Duilio, 6, 132
Duke and Duchess of York, 107

Edison, Thomas, 108, 121
Eiffel tower, 121
electric light, 21, 121
electroplating, 16
Elswick, 6, 132–3
Embankment, Victoria, 22
Emperor of the French, 86
engraving, wood, xi
Eurydice, HMS, 136
exhibitions, 110–23
Exposition Universelle (1889), 121

factory, *see* industry, manufacturing
Faraday, Michael, 92, 97
Ferndale Colliery, 12
firefighting, 28, 100

Gas, Light and Coke Company, 20
Ghost, Pepper's, 95
Gilbert, Sir John, x
Gladstone, William, 8, 18
Gloucester, 51
Gooch, Sir Daniel, 40
Great Britain, SS, 42, 43
Great Eastern, The, 39–40, 44, 47
Great Exhibition (1851), 90, 110, 112, 114
Great Western, SS, 42, 47
Great Western Railway, 36, 51, 74
Grimthorpe, Lord, *see* Denison
Grubb, Sir Howard, 104
gunnery, 130–3
Gurney, Samuel, 25

health, *see* hospitals, medicine, sewers, water supply
Hood, HMS, 140
Hjorth, Søren, 110

hospitals: London, xii; Netley, 124
Hotel, Midland Grand, 68
Huxley, T.H., 94, 97

Illustrated London News, viii–xi
Impulsoria, 59
induction coil, ii
Industrial Revolution, vii, 18
industry: brewing, 4; coal, 11–13; iron and steel, 8, 16, 116, 126, 133; manufacturing, 14–16, 18, 117; mining, 10–13; munitions, 126, 130–3
Ingram, Herbert, x
International Electric Exhibition (1882), 121
International Exhibition (1862), 116–17

Jackson, Mason, vii

Kelk, Sir John, 116
Koch, Robert, 31

Lever, W.H., 18
'Lady with the lamp', 30
lighthouse, 50
lighting, 21, 50, 121, 136
Lister, Joseph, 32
Liverpool, 24
Loudoun Castle, SS, 4

Manchester Ship Canal, 34
McConnell, J.E., 58, 80
mechanical handling, 11
Medhurst, George, 53
medicine, xii, 30–3, 124
Merrimack, 135
Metropolitan Fire Brigade, 28
Metropolitan Underground Railway, 74
mining, 10–12
motor-car, 89
Muybridge, Eadweard, 106

Newcastle upon Tyne, 6, 15, 133
Niagara, 66
Niagara, USS, 38
Nightingale, Florence, 30
Novgorod, 136

Owen, Sir Richard, 94

Paris, 121, 129
Pasteur, Louis, 32
Paxton, Joseph, 112, 114
Pepper, 'Professor', ii, 95
phonograph, 108, 121
photography, 105–6
pistolgram, 105
Platt, John, 117
Preece, Sir William, 21, 41
Prince Consort, 110, 114
Prince Imperial, x, 86
Prince of Wales, 4, 16–17
Princess of Wales, 16–17
Pullman, 78
pumping station, 26
Punch, 25, 112

railways 15, 36, 51–79, 132; atmospheric, 22, 52; compressed air, 70; pneumatic, 53; stations, 52, 62, 64, 68; underground, 22, 74
Roebling, John, 66
Röntgen, Wilhelm, 107
rotogravure, xi
Royal Institution, ii, 92, 94, 97
Royal Observatory, 37, 100, 102
Royal Polytechnic, ii, 95
Royal Society, 32, 85
Russell, John Scott, 44

Science Museum viii–ix, 96, 110, 128
Scott, Sir George Gilbert, 68
sewers, 22, 26
Shaw, Sir Eyre Massey, 28
Sheffield, 16, 126
ships, 38–40, 42–9, 126, 130, 132–6, 140
Sibthorp, Colonel, 114
Smithfield Cattle Show, 3
Society of Arts, 114, 117
South Kensington Museum, 94, 96, 110
speaking tube, 100
steam: carriage, 80; engine, 28; horse, 2; locomotive, 58; roller, 88
Stephenson, George, 15, 51, 54, 63
Stephenson, Robert, 6, 15, 54–5

tea trade, 4
telegraph, 36–40
telephone, 41, 121
telescope, 102–4
Telford, Thomas, 55, 82
textiles, manufacture, 18, 117
Thames, River, 22, 26
Thomson, William, 39
time ball, 37
time measurement, 37, 102
tramcar, 85
transport: air, 129; land, 51–89; water, 42–9; *see also* railways, ships, etc.
Tsar, The, 130
tricycle, 87
tuberculosis, 31
tunnels: Mont Cenis, 70; St Gotthard, 70
Tyndall, John, 92, 97

velocipede, 86
Viaduct, Holborn Valley, 84
Victoria, Queen, xii, 34, 39, 71, 90, 105, 110, 124
Victoria Station, 62

Warrington, 18
water supply, 24–6
Wheatstone, Charles, 36
'whippers', 11
Wolseley, Sir Garnet, 139
wood engraving, xi
Woolwich Arsenal, 130
Wordsworth, William, 8

X-rays, 107

zoöpraxiscope, 106